SCIENCE AS A CULTURAL FORCE

SCIENCE AS A CULTURAL FORCE

Edited with an Introduction

by

HARRY WOOLF

The Shell Companies Foundation Lectures

THE JOHNS HOPKINS PRESS
Baltimore, Maryland

Printed in the United States of America
Library of Congress Catalog Card Number 64-25077

Table of Contents

Biographical Notes

HARRY WOOLF is Willis K. Shepard Professor of the History of Science at The Johns Hopkins University and Chairman of the newly created Department of the History of Science. A former editor of *ISIS*, he is author of *Transits of Venus: A Study in the Organization and Practice of Eighteenth-Century Science* and editor of *Quantification: A History of the Meaning of Measurement in the Natural and Social Sciences.*

JAMES R. KILLIAN, JR., became chairman of the corporation of the Massachusetts Institute of Technology in 1959, after nearly ten years as president. His administration directed a sharp expansion of the Institute's educational and research programs and laid increased emphasis on the humanities and social sciences in educating scientists and engineers. From 1957 to 1959 Dr. Killian served as Special Assistant for Science and Technology to President Dwight D. Eisenhower.

JEROME B. WIESNER, Dean of the School of Science at the Massachusetts Institute of Technology, has been a leader in the development of communications. He was appointed to his present position at M.I.T. after service from 1961 through 1963 as Special Assistant for Science and Technology to President John F. Kennedy.

Much of his effort has been devoted to the search for methods of arms control, and he has been active in the technology of national defense. In addition to his post as Special

Assistant for Science and Technology to the President, he also served as Director of the Office of Science and Technology.

MICHAEL POLANYI began conducting research in physics and chemistry, first as a member of the Kaiser Wilhelm Institute for Physical Chemistry, then as Professor of Physical Chemistry in the University of Manchester. He entered still another new field in 1948, when he exchanged the chair of Physical Chemistry at Manchester for the chair of Social Studies in the same university. His research and writing since have cut across the conventional boundaries of academic subjects.

After his retirement from Manchester in 1958 he became a Fellow of Merton College, Oxford, from 1959 to 1961 and is presently a Fellow at Duke University. His most recent publications are *The Study of Man* and *Beyond Nihilism.*

GERALD HOLTON, Professor of Physics at Harvard University, combines experimental research and teaching with a broad interest in the history and philosophy of science. In his experimental research work, Dr. Holton has been investigating the properties of materials under high pressure, particularly relaxation processes in liquids. He is the author of two widely used books: *Introduction to Concepts and Theories in Physical Science* and *Foundations of Modern Physical Science.*

In 1960–61 Dr. Holton was National Science Foundation Faculty Fellow at the University of Paris, and in the spring of 1962 he was an exchange professor at Leningrad University. He was on sabbatical leave from Harvard during 1963–64 as a member of the Institute for Advanced Study in Princeton, New Jersey.

SCIENCE AS A CULTURAL FORCE

I

Science in Society

HARRY WOOLF

THE CONCEPT OF a university as an isolated community of scholars is as remote from contemporary reality as the recently fashionable quip which encapsulated the academic world as a group of individuals held together by a common grievance over parking space. The persistent ideal of the former and the bare truth revealed by the wit of the latter do depict, however, some of the troubling aspects of the life of learning in our time.

The common scholarly language, the unique orientation and goal of the early university community as a single enterprise at one with its general environment, even if separated from it by professional, specialized activities, is nonexistent today. Society's practical problems increasingly thrust their way into study and laboratory, insistently narrowing the distance between theory and application, demanding that academic intelligence come into the market place, indeed, offering the tempting, obvious returns for such action. Defense contracts, foundation awards, and the ever-broadening use of the professor as adviser, witness, or consultant indicate what other instruments of judgment have long measured—the destruction of university isolation and the rising involvement of its activities with those of society at large. The sciences have led in this development, with their promise of immediate gain, but more importantly, because of their capacities to gather, to order and to retrieve for utilitarian or philosophical decision the varied facts that populate the many worlds of modern existence.

Since the new philosophy of seventeenth-century discourse came to cast in doubt all that men had held as certain before—a revolutionary achievement which historical judgment conveniently identifies with the heliocentricity of Copernicus and the new physics of Galileo—the sciences, as exponents of certain essential themes and proponents of certain precise methods, have propelled themselves to the frontiers of humanity and held sway there, for weal or woe, as the very cutting edge of civilization. To free itself from the medieval embrace, to first recover and afterward crack the chrysalis of Greek thought in which so much of its own destiny and character were formed, science attempted continuously to reject the individual elements of private person or social circumstance, and to count as its triumph the naked, neutral achievement of mathematical analysis—indeed, often to reserve for this accomplishment alone the sacred word *pure*.

Nevertheless, our evolving understanding of the historical and sociological forces which mold the raw psychology of man into savant or saint, decent citizen or habitual criminal have taught us to look for the origin and changing character of science, even in its most abstruce forms, in the vagaries of time, person, and place as well as within its own internal dynamics, once under way. Such a view reveals clearly that science is not only a handmade, cultural artifact, and not merely an imprisoned absolute which the scientist releases as he fractures the bonds of nature, but also the quintessential truth that science and the society which it helps to shape, as it is influenced by it in turn, constitute a single culture in whose broad sweep the dialogue and the vocabulary may vary dramatically, but the grammar of learning remains constant.

In recognition of this truth and because today's discourse is strained by words unknown or misunderstood, by the polarities and divisiveness generated in technical vocabularies (and not in science and technology alone!), and the apparent disappearance of values and purpose beneath a flood of technological successes, the essays in this volume were assembled. Taking science and culture in their broadest possible meaning, a group of distinguished scholars, whose lives and work have demonstrated a concern for these matters, were invited to deliver public lectures at The Johns Hopkins University on the topics which form the main chapters of this book. Although the lectures were separately given during the spring of 1964, the plans to bring them together into a single volume that would give them the character of a symposium were part of the original intention of the committee which selected the speakers and set the general frame for the series.

The interrelationship of science and government is one of the key issues of our times, a basic factor in the contemporary growth of science itself and a challenge to those concerned with free inquiry and the preservation of the democratic way. The experience of Dr. James R. Killian, Jr. in these matters, as a former science adviser to President Dwight D. Eisenhower and the present Chairman of the Board of the Massachusetts Institute of Technology, is of inestimable value. His discussion of the "knowledge industry" and the "innovation industry" reflect the mature judgments of a man who has helped move America toward that "research-reliant society" which is an essential part of his essay as it has been of his career. As governmental budgets for science mount into the billions, the regional pressures upon the federal largesse increase proportionately. But while there is no gainsaying the wisdom of multiplying our research and development establishments, with their attendent benefits in local economic growth, the spread of cyclotrons, nuclear reactors, and research institutes cannot parallel the distribution of federal funds for building highways or improving rivers and harbors. Unlike the highway system, the national pool of scientific talent does not spread equally across the nation. Dr. Killian's thoughts on these and other, related issues are of vital concern to the free world.

Closely related to Dr. Killian's comments are those by Dr. Jerome B. Wiesner, Dean of the School of Science at the Massachusetts Institute of Technology, and science adviser to the late President Kennedy. Under the general topic of "Technology and Society," Dr. Wiesner has not only reinforced the judgments of Dr. Killian with those drawn from his own experience, he has gone beyond them to discuss in detail the technological forces which shape modern society and some of the new uses to which the sophisticated instruments of research can be put. The computer, he tells us forcefully, has precipitated a major industrial, and consequently, a social revolution. Above and beyond the ability of computers to perform mathematical tasks with speed and wizardry beyond human capabilities, Dr. Wiesner emphasizes that dimension of computer application which makes possible the modeling of complicated dynamic systems. The simulation and subsequent investigation of innumerable forms of economic and social behavior make possible human, and humane, decisions in advance of their actual application to living people. The Pakistan study which Dr. Wiesner relates is a dramatic illustration of that promising potential for the "relief of man's estate." Included in Dr. Wiesner's general discussion is a proposed yearly, billion-dollar budget for research and development, but he offsets the weight which that figure seems to carry by indicating that it is "less than 20 per cent of the space budget and only 5 per cent of the presently available unused resources in the economy. . . ."

A distinguished scientist of great, international fame, Dr. Michael Polanyi chose to discuss science and philosophy under the theme of man's place in the universe. Dr. Polanyi has written and argued extensively on this subject, for he has been concerned deeply with the process by which order and coherence are extracted from our apprehension and manipulation of nature. The discovery of significant patterns in nature ultimately involves an act of personal judgment and participation that sets aside the established scientific ideal of disinterested neutrality in the face of phenomena. Final decisions in the shaping of theory are tacitly made, and theories themselves provide, as he has remarked elsewhere, the main driving forces to discovery. Major theories not only reveal the order incipient in the stuff of the world, they also point implicitly to the coherence to be found at still greater depths. In this sense a twofold truth emerges. Dr. Polanyi argues: "we know more than we can tell," and we must believe in a hierarchical cosmos. Man discovers his place in such a universe by integrating his knowledge of all below him into a meaningful array, even as he recognizes his possible emergence to a higher level of existence. This theme of personal knowledge and emergent evolution in the very bosom of science that is otherwise defined today will be rejected by many and welcomed by some, but none will deny the quest itself: to find a home for man in the universe. In this search, as in his lifelong accomplishment, Dr. Polanyi clearly exemplifies the unitary aim of this symposium just as, with equal force, he shatters the popular myth of the two cultures.

Dr. Gerald Holton's discussion of presuppositions in theory-building mirrors his established historical and philosophical interest in an analysis of the intellectual structure of science. In various publications he has attempted to sift separately the experimental, the theoretical, and the thematic elements in science, a task which has led him from a study of Johann Kepler and the philosophical origins of modern physics to a critical evaluation of what he has called the "changing allegory of motion." His effort to get at the architectonic quality and inner tonality of science is thus an attempt to render increasingly concrete those aspects of the scientific enterprise that still remain vague and amorphous.

"We find," he writes, "even among the most tough-minded philosophers and scientists a tendency to admit the necessity and existence of a non-contingent dimension in scientific work. . . ." It is in this realm of the non-contingent that certain themes assert and reassert themselves in the history of science, principles of "potency" and conservation or constancy, for example, which are not subject to common scientific analysis. Here the historian and philosopher of science and the humanist-at-large may find common ground with the scientist in the endless search for an understanding of man and nature. If scientists themselves rarely discuss these matters *ex cathedra*, it is because they have been deeply involved with the technics of science—the great formulations which mark its visible successes—because there are no final solutions to thematic problems, and because (it needs to be said) they have not, in most cases, been educated to appreciate the literature or to engage in the fascinating, fruitful dialogue that distinguishes this country of the mind where all men of learning are one.

This statement by J. Clark Maxwell is perhaps an exception to prove the rule: "discussion of questions of this kind has been going on ever since man began to reason, and to each of us, as soon as we obtain the use of our faculties, the same old questions arise as fresh as ever. They form as essential a part of the nineteenth century of our era, as that of the fifth century before it." (Quoted by C. C. Gillispie in *The Edge of Objectivity*, p. 477, and again by Dr. Holton.) Thus, for all its concern with science as such, the emphasis upon the thematic element in its structure which highlights the final essay of this volume also brings into relief that place where scientist and humanist may stand together in separate but complementary association.

Our original purpose was to encourage a discussion of science that would reveal its unfolding character as a human endeavor and assist our understanding of it as a force among us. But science is neither Aladdin nor Frankenstein, save as it reflects our own aspirations writ large or small. By dividing our discussion into the parts of science that interlace with government, technology, philosophy, and the humanities, we have sought to deal with real issues, to locate science properly in the matrix of modern culture. Although the inquiry which these papers have initiated has focused upon contemporary science, the interest of the Department of the History of Science at Johns Hopkins in acting as their local sponsor has not been entirely disinterested, for the issues raised in these essays as immediate concerns have their roots in a surprisingly distant past whose internal critical analysis constitutes a humane and continuing evaluation of science as human experience.

The success of this series would not have been possible without the willing co-operation of far too many to thank here. The committee which met to set the topics for discussion and to select the speakers consisted of Professor H. E. Hoelscher of the School of Engineering, Professor Maurice Mandelbaum of the Department of Philosophy, and Professor Carl Swanson of the Department of Biology. In addition to their help, Dr. Milton Eisenhower was kind enough to open the lecture series by introducing Dr. James R. Killian, Jr., Professor William D. McElroy introduced Dr. James B. Wiesner, and Professor Carl Christ presented Dr. Michael Polanyi. To all of these colleagues I should like to express my warmest thanks. Additional assistance, often beyond the call of their official tasks, came from Mr. Wilbert Locklin, Mr. Philip Murphey, Mr. John Synodynos, Mrs. Elizabeth Murdock, Miss Aleida Cattell, and Mrs. Sara Lee Grandes del Mazo.

The greatest thanks of all, however, is reserved for the Shell Companies Foundation. Its generous grant, given freely without restrictions of any kind, made the entire lecture series and the present volume possible. Such enlightened activity by American industry is a mark of its mature interest in the intellectual and educational issues which are the common concern of us all.

II

Toward a Research-Reliant Society: Some Observations on Government and Science

JAMES R. KILLIAN, JR.

ECONOMISTS SPEAK TODAY of the "knowledge industry." In similar argot, let me speak of the innovation industry as shorthand for "research, development, test, and evaluation" and the establishment which is responsible for these activities. After a slow growth in the United States, the innovation industry reached maturity in World War II, and its decisive contributions then to the winning of the war set the stage for the spectacular growth it has since enjoyed.

Between 1940 and 1960 scientific research and development in the United States grew at an annual rate of nearly 20 per cent.[1] In 1964, counting both public and private funds, the nation spent over 3 per cent of its gross national product for research and development, or about $20 billion. Of this total the federal government financed $15 billion of which $1.5 billion is for basic research and the rest for development and technology, including military and space programs. Eighty per cent of the federal funds

[1] Fritz Machlup, *The Production and Distribution of Knowledge in the United States* (Princeton, N. J.: Princeton University Press, 1962), p. 155.

for research and development are expended by two agencies, the Department of Defense and the National Aeronautics and Space Administration. About 400,000 scientists and engineers are engaged in research and development, of which 75 per cent are employed by industry, 11 per cent directly by government, and 12 per cent by colleges and universities. The government now supports, directly or indirectly, about 60 per cent of the nation's scientists and engineers engaged in research and development. I hasten to point out that other nations of the West, as for example Great Britain, are investing in research and development about the same percentage of their gross national product as we are.

With more money now being spent on research and development by the federal government than its total budget before Pearl Harbor, science, as Dean Price of Harvard tells us, "has become the Major Establishment in the American political system: the only set of institutions for which tax funds are appropriated almost on faith, and under concordats which protect the autonomy, if not the cloistered calm, of the laboratory."[2]

While I am not certain how to define this "establishment" and am even less certain about its relative size, it nevertheless has now become a highly visible feature of the political landscape, and as a result has been caught up in the "politics of scale." Under these conditions, it offers both members of Congress and the student of political science a new and still exotic domain to explore.

Obviously the *rate* of growth of the innovation industry we have witnessed in the last two decades cannot continue because of a limit on both dollars and men, although in my

[2] Don K. Price, "The Scientific Establishment," *Proceedings of the American Philosophical Society,* Vol. 106, No. 3 (June, 1962), p. 235.

judgment our investment in basic science should continue to grow. If we are to increase our output of doctorates in science and engineering, as all informed judgment indicates we should, basic research must grow because of its interrelation with graduate education. The maintenance and strengthening of this interrelationship should be a firm policy of the federal government.

In the period since World War II great progress has been made in providing new government mechanisms for dealing with an advancing science and technology and in augmenting the government's contribution to furthering national progress in these fields. In the last decade or so, especially, notable improvements have been made in the Executive Branch of the government in using scientific advice in policy-making and in the co-ordination of the government's own scientific activities. The latest of these important advances is the Office of Science and Technology. Even with these useful new mechanisms and with an increasing scientific competence in government, there remains a growing concern as to whether the process of adaptation to new requirements has been adequate. This concern relates both to the Executive Branch and the Legislative Branch, and many informed people, both in the community of science and in the community of public administration and legislation, feel that an intensive review of the government's science program is timely. This feeling reflects in part the greater understanding and insights of scientists and engineers who have had extensive experience in government, together with the new insights and creative ideas of political and other social scientists who have been studying this experience. The time seems right to mobilize and combine these insights and new bodies of experience and study in a comprehensive review of science and public policy. Such reviews are being undertaken in three parts, one centered in Congress and dealing with legislative matters, another centered in the Executive Branch and dealing with executive problems, and the third centered in the scientific community itself.

This period of transition and stocktaking is crucial for the future of science. In the spirit of the dialogue now taking place, let me make some personal observations about factors which seem to me to be important as we review present practices and establish future goals.

The overriding fact in all current discussions is the flourishing state of American science today. As Sir Charles Snow tells us: "It sometimes surprises Europeans to realize how much of the pure science of the West is being carried out in the United States. Curiously enough, it often surprises Americans too. At a guess, the figure is something like 80 per cent and might easily be higher."[3] Marked both by fecundity and brilliance, our science is also "full of growing." Although we do not have as many as we need, we have university centers of science and engineering unexcelled perhaps in the world. Foreign scientists leave their home countries to take advantage of the intellectual excitement, the ambiance of freedom and esteem for science they find here, the wealth of equipment, the salubrious research environment, and the absence of professional class distinctions separating science and engineering. The resulting "brain drain," or "unfavorable balance of trade" in technical personnel, as we know, has become a political issue for at least one of our allies.

[3] C. P. Snow, 110th Anniversary banquet speech, Washington University, St. Louis, Mo., February 23, 1963.

American pre-eminence has come fast. It is interesting to recall a statement by I. I. Rabi that the American journal, *The Physical Review,* was not judged to be outstandingly valuable when he studied in Germany in 1927. Ten years later, he said, it had become the leading physics journal in the world.[4]

In 1902 the economist Carl Snyder was writing that one could search the world's scientific literature in vain for references to distinguished American achievement.[5] Nearly twenty years earlier the physicist Henry A. Rowland complained that "No physicist of the first class has ever existed in this country."[6]

Today even though American science, by contrast, is flourishing, there are clouds on the horizon, especially the technological horizon. The first is the inclination to let down at a time when undiminished effort can bring great returns. At the risk of indulging in over-confidence, I venture the conclusion that we may be only at the beginning of unexampled scientific and engineering achievement. With the future so promising, this is not the time to relax our scientific effort or for timid talk about having reached some kind of ceiling in our upsurge of scientific and technological productivity. At this juncture the role of the federal government is crucial. What policies it follows in support and encouragement can largely determine whether our progress is retarded or advanced.

[4] I. I. Rabi, Transcription of lecture at M.I.T. (unpublished).

[5] Carl Snyder, "America's Inferior Position in the Scientific World," *North American Review,* Vol. CLXXIV, No. DXLII (Jan. 1902), pp. 60–72.

[6] Henry A. Rowland, "A Plea for Pure Science" [address as Vice President of Section B of the American Association for the Advancement of Science, Minneapolis, August 15, 1883], *The Physical Papers of Henry Augustus Rowland* (Baltimore: The Johns Hopkins Press, 1902), p. 160.

The second hazard is that the scale of our present effort and our current high confidence may obscure weaknesses in our program. We may not be as strong tomorrow as we appear to be today. Indeed in some quarters, the post-Sputnik sense of urgency having subsided, we seem again to be growing smug about our strength. I have an uneasy feeling that the pendulum of our concern has now swung back to the contented self-satisfaction which lulled and charmed us during the months prior to October, 1957. Variations in the public pulse and the political temper, both domestic and international, tend as never before to affect attitudes toward science and its support now that government is so closely coupled with science. As we examine our programs, policies, and attitudes today, we should seek to discourage these extreme oscillations in our national attitude.

Let me cite three considerations which should give us pause as we preen ourselves on our scientific and technological leadership. First, our industrial technology faces increasingly able competition from abroad. In Europe, industrial research is recapturing much of its prewar vigor, while in Japan we witness a high order of technological accomplishment out of which are coming sophisticated products, especially in electronics, which are giving American products tough competition.

In its 1964 report, released in March, 1964, the Joint Economic Committee of Congress noted "evidence that our margin of technological superiority may be diminishing as other nations step up their research and development efforts." The report makes the further observation that despite the great size of our over-all effort, our current level and allocation of research and development expenditures may be inadequate for the "sustained and rapid economic growth" we need and that "the large increase in military and space research in recent years may have created an imbalance in the allocation of our research talent leading to a shortchanging of research in many parts of the private sector."[7]

If we are not alert to the implications of such trends, we could again find ourselves taken by surprise. Nationally we seem at times to be surprise-prone. Our self-confidence sometimes impoverishes forethought. In the current environment of rapid movement, we must be prepared for the unexpected; we have no choice but to be a seat-belt society. I hope that we do not again find ourselves suddenly astonished by the gathering strength of technology abroad.

There is some evidence that the growth of research and development financed by American industry itself, expressed as a percentage of sales, has slowed down. Have the growth and policies of the government's research and development program contributed to this decrease in rate of growth? Has the impact of the federal program so increased the cost of research that industry has been unable to afford to increase its own research?

Research and development—whether privately or federally supported—which is undertaken by industry is concentrated in a few industries. In fact, over half of all industrial research in the United States is concentrated in two industrial groups: the aerospace industry and the electrical and communications industry. Three-quarters of all research and development financed by industry's own funds is done by only 200 firms. Obviously we have

[7] U.S., Congress, Joint Economic Committee, *Report on the January 1964 Economic Report of the President,* 88th Cong., 2nd Sess., 1964, Senate Rept. 931, p. 55.

not yet created the incentives and conditions in this country which would lead to a deep penetration of research and development into our industrial community.

My point is that we should not be bemused by "it-can't-happen-here" complacency. It did happen to Britain about 100 years ago. Prior to 1850 she had undisputed technological leadership, being the home of the industrial revolution. By 1870 her industry had become content with the machines it then had, however, and failed to recognize that technology requires ceaseless development; there was a consequent withering of innovation and enterprise. Meanwhile, Germany was discovering ways to exploit science systematically to advance technology. She pioneered in the development of the innovation industry, her universities became the seedbeds of new industry, and by the turn of the century she had captured world leadership in industrial technology, with the United States coming up fast on her heels.

While chiefly of academic interest now, another instructive example of failure to use important scientific and technological advances for social purposes appears in the ancient history of China. Before 1000 A.D., she led the world in technical developments, but as Professor William H. McNeill reports in *The Rise of the West,* "The full and reckless exploitation of these inventions was reserved for the looser, less-ordered society of Western Europe, where no overarching bureaucracy and no unchallengeable social hierarchy inhibited their revolutionary application."[8] With our massive government support, we may be creating the overarching bureaucracy, but the social hierarchy is fortunately absent.

[8] William H. McNeill, *The Rise of the West: A History of the Human Community* (Chicago: University of Chicago Press, 1963), p. 469.

A second consideration is the importance of maintaining undiminished a steady input of creative ideas into our advanced weapons technology. We may fervently hope that the cold war is moderating and that the enormous share of our intellectual energies it has commanded can be reduced. Under these conditions of enhanced hopefulness and progress, however, we may find greater difficulty in maintaining a continued high creative input into our military technology. Along with our industrial technology, our military technology, to be superior and even adequate, requires unceasing innovation and advance. There has been no partisanship in adhering to a national military policy that calls for us to maintain a margin of superiority as the best means for preserving the peace and making progress toward safe arms limitation.

We must also be aware of the possibility today that the government, in a laudable program to achieve better management and supervision of research and development, may be unwittingly imposing constraints on the creative process. In the late forties and well into the fifties, we had a revolution in weapons technology through a long input of innovative ideas. But we must now take care to avoid reducing the input of new ideas into military technology by an overinstitutionalization of the relation of science to government. It is also important that we not permit theoretical analysis and management concepts, important as they are, to impose too much order and inhibition on the creative process, which tends to flourish best under conditions of mild disorder. Our efforts to supervise and monitor the innovation industry in the years ahead—to make it more efficient—should not be permitted to have the reverse effect of slowing it down. These are subtle matters, and they warrant careful attention.

A third consideration which should give us concern lies in the domain of education. The quality of our education in the sciences still needs to be improved.

In their book, *Education, Manpower, and Economic Growth,* Frederick Harbison and Charles A. Myers develop some significant indices comparing the manpower resources of countries in various stages of development. One of their comparisons is the distribution of students between science and technology on the one hand, and humanities, law, and the arts on the other. In the percentage of its students studying science and technology, the United States stands substantially below the mean of sixteen advanced countries. Sweden, Israel, West Germany, Finland, Russia, Canada, France, United Kingdom, Belgium, Netherlands, and Australia, all have a higher percentage than the United States studying science or engineering. The Soviet percentage is twice that of the United States. We hear sometimes that we are overemphasizing science, but this comparison certainly does not support this contention. We actually face a problem of too few students electing science and engineering to meet our national needs. To cite but one need: The President's Science Advisory Committee has recommended that we at least double our production of doctorates in science, engineering, and mathematics by 1970.

In recent years the government initiated a number of support programs to meet these challenges. The Defense Education Act of 1958 has bettered science teaching resources in precollege schools, as indeed it has strengthened language instruction. The program of the National Science Foundation for strengthening science and engineering at the college and university level has been profoundly important. The recent education act passed by Congress will help to meet the vast building needs of our rapidly growing system of higher education.

Federal support of research in educational institutions has served greatly to strengthen our universities as well as American science. This federal support has been generally well conceived and administered and, in turn, it has been in the main well used by the universities.

Many people, especially representatives of disciplines other than science in the universities, are critical of the current emphasis on science—and of the government concentration of its support in the domain of science. They feel that the large government support it has been receiving has created a serious distortion in our scholarship and culture and that the humanities have been neglected.

President Clark Kerr has said some wise things about this. "The essence of balance," he emphasized, "is to match support with intellectual creativity of subject fields; with the needs for skills at the highest level; with the kinds of expert service that society most requires."[9] There have been periods when the humanities possessed the privilege of the fair-haired boy, and they may do so again, but today the tide is running for science because the nation needs strong science, and we should not feel our government is distorting the pattern by responding to these conditions.

We also hear debate about imbalances within science, especially concern about overemphasis on high-energy physics, but it may well be right to lay our bets on a robust specialty that is currently so attractive to many of our gifted scientific minds. Their sense of what is important may be the best guide available.

[9] Clark Kerr, *The Uses of the University* (Cambridge, Mass.: Harvard University Press, 1963), p. 114.

Prophecy is dangerous, but it would appear that the social sciences, invigorated as they have been by methods from the physical sciences and mathematics, may be headed for a period of boom and growing support. Perhaps both they and the humanities, challenged by the flourishing state of science, may enhance their vigor and support and thus correct any imbalances which now exist. While our humanities are strong today, it is essential to the quality of our society, including its science, that they be stronger still, and that they have adequate support. The arts in America seem to be already well down that road. At this juncture we also need a great manifesto for the humanities with a power to move men comparable to Vannevar Bush's *Science: The Endless Frontier.*

One of the more widespread criticisms of government sponsored research in the universities is that it takes priority over teaching. Teaching, it is said, has become a poor relation of research, not to speak of advisory and consulting activities outside of the university. I think there is much exaggeration in this criticism. While unquestionably there has been some neglect, in some places, of undergraduate teaching by university professors, only in part can this neglect be charged to sponsored research and federal funds. Actually, there is ample evidence that basic research can and does invigorate undergraduate teaching. Our universities have a superb opportunity to encourage the penetration of research into the undergraduate domain to enrich the teaching and to provide an atmosphere of creative vigor for both students and staff.

The universities have gone through a period when great efforts have been devoted to strengthening graduate study. We are now entering another cycle when creative energies are being applied to curriculum reform and to the teaching process, not only in the undergraduate schools in the universities but in precollege schools as well. With strong support from the National Science Foundation, professors of science in our universities have rolled up their sleeves and with revolutionary effect gone to work on preparing new and better materials and methods for the teaching of science from elementary school to the graduate school. This creative approach to the teaching process on the part of leading research scholars is now spreading, and is an eloquent counter to any neglect of teaching which has been evident in recent years. The curriculum reform programs in mathematics, physics, chemistry, and biology not only are giving a new depth and maturity to American education; they also are fine examples of a productive interaction between government and the civilian scientific community.

These general considerations and national requirements should be clearly in view as we re-examine our national science policies, especially those affecting government participation. In addition there are other important questions which are currently being examined and debated. Let me note three of these.

First, there is the vital requirement that excellence be the touchstone of our national research planning. As the President's Science Advisory Committee has emphasized, "In science the excellent is not just better than the ordinary; it is almost all that matters. It is therefore fundamental that this country should energetically sustain and strongly reinforce first-rate work where it now exists."[10] In the support of research, adherence to this policy by gov-

[10] U.S., President's Science Advisory Committee, "Scientific Progress, The Universities, and The Federal Government," The White House, Washington, D.C., November 15, 1960, p. 28.

ernment agencies has been a major factor in giving U.S. science world leadership. It was the late Abraham Flexner, I think, who once remarked that getting education is like getting measles; you have to go where the measles is. So it is in getting good research results.

While continuing to reinforce existing centers of strength, we must also create new ones. We need more graduate schools of science and engineering as good as the best we now have. We achieve these, not by diminishing the strength or support of the existing great centers of strength but by encouraging others to develop. The program of institutional grants initiated by the National Science Foundation is one way of beginning to build new centers of strength and to achieve a wider diffusion of first-rank scientists and engineers.

This is not done by building on weakness. It is done rather by identifying those institutions which have shown the initiative and mobilized the support to strengthen themselves. More constituencies—communities and states—should determinedly set about to strengthen their institutions to the point where sources of funds, public and private, can justify helping them grow stronger still.

The widespread recognition that strong centers of research and development, especially as parts of a university complex, enhance the prosperity of a region or community has led to increasingly fierce geographical competition for federal research and development funds. Congressmen have been understandably responsive to these local pressures, and there is an increasing chorus of voices critical of the paucity of research support in some regions and communities. The way that this conflict between regional needs and quality is resolved will be a test of how well our political system can deal with science.

I agree with Dr. Glenn Seaborg when he says that "we must not let our national support of science and technology degenerate to the point where no state—no Congressional district—is complete without a Post Office, a reclamation project and a new science laboratory. Any such program, however, should be both soundly conceived and wisely administered if we are to build new centers of excellence in new geographical areas without tearing down or undermining other centers that have already achieved and sustained excellence."[11] Still, we must devise plans and policies for creating new centers of strength, and we must do it promptly.

I turn next to manpower utilization, one of the areas most requiring attention now that we are sponsoring such vast research and development efforts. The way scientists and engineers deploy themselves and are utilized is more important than the way we allot our research and development dollar, although the two, of course, are linked.

About 60 per cent of all of the scientists and engineers in research and development in the United States are working wholly or in part on projects or programs financed by the federal government.

Under these conditions, it is obviously important that government must try to avoid policies or procedures which may lead to inefficient deployment, building up one area at the expense of another, stock-piling, and so on. When considering the launching of big programs requiring large-scale research and development, the government, in advance, should count the cost in manpower. We now are in an era when government decisions can affect small armies of scientists and engineers.

[11] Glenn T. Seaborg, The Harrelson Lecture, University of North Carolina, March 11, 1964 (mimeographed copy).

We urgently need more information about manpower utilization. So far we have had to make decisions based on hunches, intuition, or fragmentary data.

In this random list of problem areas, I come finally to the problem of achieving good management and good advice in all institutions, public and private, of our massive innovation industry, but especially in government. One wag has likened our federal management of scientific programs "to a ship with a thousand helms all connected to one rudder with rubber bands"—a pluralistic system that has not been without merit!

The need for good management and sound advisory processes becomes the more urgent in this period of transition when the rate of growth of expenditures will probably be slowed and the problem of priorities will become crucial.

In the government, these choices will have to be made in the end by both executive and legislative processes, but great responsibility must inevitably fall upon the appointed officers, the science administrators, and their advisers. Great progress has been made in recent years in creating the necessary posts, such as Assistant Secretaries for Research and Development, and generally upgrading the administrative competence available for handling the massive government program, but much remains to be done.

The government faces a hard struggle to recruit competent technical, supervisory, and managerial talent. It constantly is up against a dangerous weakening of its management strength, largely because pay scales and personnel policies are inadequate, and technically competent administrators are hard to attract and hold. Science administrators, managers who possess a combination of technological mastery and administrative skill, project engineers, and research directors—all these are extremely scarce today. More and better ones can do much to improve the federal as well as the industrial research and development programs.

The advisory committees and other mechanisms drawing upon the resources of the nation's scientific community have served the government brilliantly in recent years and will continue to do so. They are subject, however, to constraints. Some of these arise out of the inevitable institutionalization that comes with time. These constraints appear in both government and non-government scientific activities. Scientific advisory procedures that in the beginning were effectively informal tend to become more formalized and consequently less uninhibitedly creative, especially as programs grow larger.

So great is the government's reliance on these advisory mechanisms that they have been asked to carry increasing responsibilities for monitoring and appraising government programs, for advice on the selection of projects, and for responding to the important needs of Congress. We need to give careful attention to inventing ways whereby government advisers can perform these monitoring and appraisal responsibilities and not be leached of innovation. They must continue to be a prime source of creative ideas. An occasional *ad hoc* group working full time for a limited period, such as the Technological Capabilities Panel appointed by President Eisenhower in 1954 to review our military technology, is one way to introduce fresh ideas. Advisory committees should be rotated and full-time summer study panels further utilized. In all of the many relationships which the government has with research establishments and other sources of creative scientific activity it must be recognized that while the government's interests must be carefully protected by proper supervision and control, such supervision and control, not skillfully handled, can reduce the productivity of scientists and engineers.

There is perhaps no more perplexing problem for government administrators and advisors alike than that of determining priorities and levels of spending in scientific undertakings. How much money should be allotted to basic research and how much to applied research, how much to big science and how much to little, how much to space exploration and how much to high-energy physics? I remember President Eisenhower's on several occasions asking me, or his Science Advisory Committee, to suggest a way for rationally setting a level of funds for basic research, this being in response to recommendations by his advisers that its support be increased. I do not think we were ever able to give an administratively useful suggestion other than to recommend that the limited number of genuinely creative people in pure science be adequately supported. Soviet scientists visiting this country report that they, too, are vexed by decisions on priorities, and I have just read an article on science and government in Great Britain which reports that it has no effective machinery for determining priorities. We are not alone in our perplexity.

Many suggestions have been made. Some years ago, James B. Conant advocated the development of a tradition of quasi-judicial review, including a form of adversary proceeding, in handling government technical programs, especially in the weapons field. "When a question comes up to be settled," he suggested, "one or more referees might hear the arguments pro and con. If there were no contrary arguments, some technical expert should be appointed to speak on behalf of the taxpayer *against* the proposed research and development. Then adequate briefs of the two or more sides should be prepared (*not* a compromise committee report)." Conant went on to emphasize that every citizen is a "party to an enormous new enterprise. His government has gone into the research and development business on a scale totally different from anything seen in the past. . . . Consequences of tremendous significance in terms of survival may hang on the way this work is carried on," and "the waste of enormous sums of money could threaten the soundness of our economy."[12]

I do not think that Mr. Conant's proposal has ever been tried in a formal way. It is fascinating, though idle, to speculate on what would have been the outcome if his quasi-judicial procedure had been employed in reaching a decision in regard to the lunar landing project, the proposed B-70, or, for that matter, the MURA high-energy accelerator proposals from the Midwest, the Stanford Linear Accelerator, the MOHOLE program, or the proposed supersonic transport.

Actually the President's Science Advisory Committee and its *ad hoc* panels have made available to the President, to assist him in reaching decisions, a process of judicial review that is remarkably free of the departmental biases so natural in government. In decisions relating to big science it may be important for society itself to have a say, along with the technical experts, as Alvin Weinberg has suggested. He has also advocated having as a member of review panels dealing with big science a representative

[12] James B. Conant, *Science and Common Sense* (New Haven, Conn.: Yale University Press, 1951), pp. 337, 338.

of a scientific discipline not directly involved. In formulating a recommendation on whether to build a costly high-energy accelerator, for example, it may be equitable to have scientists other than physicists participate so that the over-all interests of science can be reflected in the decision.

In seeking to determine policies and levels of support for basic research, Congress, I venture to suggest, should appraise the procedures, the methods, and the quality of management used by the federal agencies that recommend appropriations for basic research. When these appraisals are on the plus side, it should feel reasonably comfortable about the projects and the people selected to receive support. Congress itself cannot easily select these people and projects, but it can expect a selection procedure in the executive agencies in which it and the country has confidence—a procedure which requires ample justification on the part of claimants for funds. Except when questions of big machines such as research reactors or high-energy accelerators are proposed, the problem of priorities is not so acute in the domain of basic research. When big machines are involved, however, costing as they may many millions of dollars, the problem of priorities becomes very difficult indeed, and the decision-making process must rest perhaps not only upon very sound scientific judgment but upon a representative judgment that gives voice to those fields of science not involved in the proposed big machines.

I have noted that top-level science advisers have been of added usefulness because they ride above the interdepartmental conflicts in government. Three Presidents, to my knowledge, have turned to their advisory committees for objective advice on matters over which groups in government were vigorously and sometimes bitterly contending. These science advisory committees have also been apolitical. I recall President Eisenhower's once remarking that he felt the political alignment (Republican or Democrat) of the members of his Science Advisory Committee to be irrelevant to their responsibilities. In my experience I have witnessed no evidence of partisan bias in this top-level advisory process.

This is not to advocate that scientists should be politically neutral. We need more scientists actually in politics. Historically the scientist in government as scientist has sought to insulate himself and his organization from partisan political pressures. He also fears an overarching bureaucracy. Many scientists have for this reason opposed a Department of Science. They find safety in decentralized decisions and diverse sources of support. Accepting these concerns about political influence and bureaucracy, we must at the same time recognize that current trends in government resulting from the problem of priorities and the political pressures for geographical distribution of government funds presage more politics in science and thus the need for more scientists willing and able to work effectively in the political arena.

The expert, however objective he tries to be in his advice, always runs the hazard of being caught up in political judgments as he seeks to provide scientific advice. Robert Gilpin, in his book, *American Scientists and Nuclear Weapons Policy,* describes the difficult middle ground occupied by the science adviser between "the realm of science, or *what is,* and the realm of policy, or *what is to be done.*" Mr. Gilpin concludes:

> . . . that both scientists and political leaders have acted as if it were possible to make a clear delineation in policy formulation between the political and the technical realms. . . . This simplistic view of the scien-

> tist's role as adviser has created expectations which the scientist cannot fulfill. He is expected by political leadership, fellow scientists, and his own conscience to render only objective technical advice. As a consequence . . ., scientists have been assigned many apparently "technical" tasks whose performance has required a political skill far beyond their competence; scientists and political leaders have failed to realize the nature of the non-technical assumptions underlying the scientist's advice; and scientists have charged one another with intellectual dishonesty when they have disagreed strongly with advice which has been given.[13]

I cannot accept the conclusion that it is essentially impossible for the adviser to render objective technical advice in a political context, but I recognize that advisers have not always been able to be objective. The format given to the Science Advisory Committee and the directive to the Special Assistant to the President for Science and Technology in 1957 did much to reduce the possibilities of biased or personalized advice. Measures were taken to avoid the problem that Snow discussed in his story of the Lindemann-Tizard controversy in Great Britain. It was stipulated, for example, that the Science Advisory Committee had full freedom to select its own chairman and to go direct to the President if it disagreed with the Special Assistant on important matters. In its operations the President's Science Advisory Committee, as those of us know who have had experience with it, has been extraordinarily independent and outspoken and free of domination by any one person.

[13] Robert Gilpin, *American Scientists and Nuclear Weapons Policy* (Princeton, N.J.: Princeton University Press, 1962), pp. 16–18.

Still one must also emphasize, in looking generally at the problem of the expert in government, that when political considerations are intertwined with uncertain, controversial technical questions, such as those relating to a nuclear test ban, it must be expected that the expert, being human, may have difficulty separating his technical from his personal views, and the policy-maker should understand this. At the same time, it is of vital importance that the policy-maker himself not fall into the error of expecting advice to support a particular view or policy. One of the frightening aspects of the Oppenheimer case was the fear it created, although not borne out in subsequent experience, that technical advice, when not in support of some current military or political policy, might be condemned.

In recent years the political scientists have identified various ways in which scientists have modified our governmental structure or processes. One is the creation of the influential advisory process I have been discussing. A variation is embodied in the Board of the National Science Foundation, which is made up of people from outside government serving part-time and is designed to minimize political control. Most striking are the mutations which appeared during World War II when the National Defense Research Committee and the Office of Scientific Research and Development invented a contracting procedure which engaged private institutions to undertake government business. What Price has called "federalism by contract" has spread, and we now see an extraordinary array of arrangements through which government contracts for advice, policy, research, analysis, and a host of other services. And when the government could find no private institution with which it could contract for certain kinds of services, it has deliberately created private institutions, the so-called not-for-profit corporations, such as RAND and the Institute for Defense Analyses.

Largely devised by scientists, this arrangement has given the government deep roots into the scientific community, has countered tendencies toward a centralized bureaucracy, and has kept a great body of scientists engaged in work for the government free of political commitments and outside the discipline of the government's administrative organization. This has been a successful procedure in part because the work of scientific advisers has been mainly situational; they have been preoccupied with appraisal and problem-solving. They have not usually been in the mainstream of major political policy-making and action.

Some say the scientists are politically naïve, but the record says that they have been extraordinarily inventive and successful in preserving the freedom of science, especially its freedom from bureaucracy. They have also been highly responsive to requests to serve the government as advisers; they have generally been content to exercise their scientific competence without seeking to exercise political power. I am sure, however, that the methods which the government now employs to use scientists increasingly will require, if these methods are to continue to be successful, strong scientific administrators in government, fully accountable politically and working within the normal disciplines and constraints which must apply.

I see other special qualities of the scientist which give him a characteristic impact on the political process. He is usually critical of the status quo. He tends to be restless, probing, and unimpressed by tradition, rules, or hierarchy.

Living as he does on the intellectual frontier, he tends to have some of the attitudes and motivations of those of our ancestors who lived on the geographical frontier. Most characteristic of all is his instinct for innovation, his tropism for change.

This desire to improve things sometimes runs counter to orderly administration in government, and tends to make the work of the career administrator harder. And sometimes the scientist, preoccupied with improving the trees, fails to see the forest, politically speaking.

Altogether, though, the record is encouraging: the scientist has made an extraordinary adaptation to government and the government to him, with great benefits to the nation. This success has given a new significance to the role of the expert in government and is full of instructive experience as to how the expert can best serve the policy-maker.

Let me stress in this connection the vital role which the university and the university-scientist must play in maintaining an effective relationship between government and science. The university is an ideal place for the government to recruit scholars who are clearly free of serious conflicts of interests. This is not to say that conflicts of interests do not exist in the academic domain, or that scientists in other sectors do not avoid conflicts. They do, but no other environment has the opportunity and is expected by society to be so free as the university environment. It is of the greatest importance that the universities protect this freedom and objectivity, that they not compromise it by entering into any kind of relationship, whether it be with industry, government, or elsewhere, that makes it difficult for their faculties to be independent in judgment. This will not be easy in the future, but it will be vital to the integrity of the great structure we have erected in the domain of government and science.

A great body of scientists uncommitted, independent, and scrupulously objective, is the best insurance that there will be no abuses of the profound public responsibilities scientists now carry.

Implicit in the work and impact of this scientific establishment I have been examining with you is the growing national emphasis on research as a way of life. This emphasis is not confined to science. In almost every aspect of our national life we rely more and more on knowledge-seeking, problem-solving techniques. Almost every activity is invigorated by innovation. We are becoming a research-dependent society.

The innovation industry works in double harness with the knowledge industry, and the two together have become one of the principal energizers of our society, shaping its contours, constantly renewing its vitality, and making the present obsolete.

This is why so much importance attaches to the present review of science and public policy. The wisdom we bring to this review has importance beyond the boundaries of science.

III

Technology and Society

JEROME B. WIESNER

TECHNOLOGY IN THE mid-twentieth century is in a curious state. There is no doubt that it is the most dominant force in modern life. It is the engine that propels modern society, and for this it is widely acclaimed, but at the same time it has become the object of much critical examination by experts and laymen alike.

Technological activities—research and development—are among the nation's fastest growing enterprises and public funds provide much of the support for them. Widespread understanding of the importance for such support is, therefore, very desirable. During the past two decades, when the primary motivation for most of the expenditures was need for national defense, the research and development activities were becoming costly enough to attract broad public attention. The country was prepared to provide large sums of money for this purpose without too searching an inquiry. In fact, the secrecy restrictions applied to military research and development during the 1950's made serious public discussion impossible. In recent years, because of the maturity of new military technology based on rockets, computers, and thermonuclear bombs, military research and development expenditures have leveled off, and new projects receive much more scrutiny, as do non-military projects, such as the space effort, oceanographic research, the supersonic transport development, and a myriad of other non-military programs.

Fortunately for the country, the scientific and technical results of the military-sponsored research and development undertaken during the 1940 to 1960 period had direct and broad application to the civilian economy. In fact it is doubtful whether the modern aircraft, the jet engine, the electronic computer, and the many new materials in daily use would exist today if they had not been developed for military use. Certainly our recent economic growth was assisted in a major way by the development of these things. Today's military technology does not contribute so directly to our civilian economy, and consequently, we now need to undertake programs specifically devised to maintain a rapid rate of technological development in fields directly related to the civilian economy. Undoubtedly much of this support can come from private funds, but in many important areas incentives for private investment are inadequate. This may be so because of the long-term scale of the work, or because, as in the case of fundamental research, no special advantage is obtained by doing the work, or because the activity may be too costly for private industry to support. The development of nuclear power for civilian use is an example of a project with too long a time scale to be attractive to private industry. The supersonic transport aircraft is an example of a technological undertaking too costly for private financing.

At the present time two congressional committees are making detailed investigations of federal research and development activities. Their efforts appear to be largely complementary and both of them are obviously dedicated to doing a constructive job. The Select Committee, chaired by Congressman Carl Elliot of Alabama, is concentrating on management effectiveness in the federal programs. The second committee, a Subcommittee of the House Committee on Science and Astronautics, chaired by Congressman Emilio Daddario, is primarily concerned with processes by which Congress makes decisions on research and development programs. They are particularly interested in the criteria that should be applied in determining whether or not public investments should be made in technical projects.

Finally, there is widespread concern about the adequacy of our supply of technical manpower—scientists and engineers—to staff our national commitments, and interest in the quality of their education. Some of this interest was generated by the disclosure that the Soviet Union is training twice as many people in these fields as we, but concern also stems from difficulties encountered in carrying out some of the major technical projects.

I would like to clear up a continuing source of misunderstanding that appears in discussions of technological development; the ambiguity that exists between technology and scientific research, their objectives and their methods. This confusion is by no means surprising since modern technology has become highly dependent upon basic scientific knowledge for much of its progress. In turn, scientific research in many fields is only possible because of the elaborate and sensitive tools that technology has made possible. The vast and powerful particle accelerators, the electron microscope with which to explore the world of cells and viruses, and the electronic computer to calculate problems which only a few years ago were beyond the scope of human comprehension, are but three of a large number of scientific tools which have extended enormously our ability to measure, observe, and understand the world around us.

This close alliance between science and technology, though relatively new, is so complete that the average person, and indeed many scientists and engineers as well, fail to distinguish any difference between them.

In the beginning, technology did not depend upon science. The inventions that provided the basis for the industrial revolution—the steam engine, the loom, the lathe, and many other machines—were invented by practical men and based upon art, observation, and common sense. In the first stage of industrialization man was exercising his ingenuity in the exploitation of the things he found around him. The factory with its power machines, its use of unskilled or semi-skilled labor doing simple repetitive operations, the utilization of raw materials like iron, coal, copper, etc., improvements in transportation growing from the development of the railroad and the steamboat are all examples of this inventiveness. Most important, of course, was that with the introduction of machines he had begun a continuing process of extending human capabilities, first by augmenting muscle power through harnessing the almost limitless energy sources found in nature, later by speeding communications by electrical means, and most recently by augmenting mental activities by the introduction of computing machines to replace human effort in menial, repetitive activities and to assist in difficult or lengthy calculations.

The fact that scientific research had little or no effect on early technology does not mean that scientists did not exist or were not working. They did and were, and during the period of the industrial revolution the foundations were laid for modern physics, chemistry, and biology.

However, it was not until the middle of the nineteenth century that extensive practical use was made of the accumulating scientific knowledge. Only then did men begin to exploit the available knowledge of chemistry and electricity for useful purposes. Chemists learned to synthesize organic materials and set up research laboratories for obtaining the new knowledge required to meet their applied objectives. It was in the field of chemistry that research methods were first applied in a systematic manner to develop new products. The application of electricity was more haphazard in the beginning. The scientific observations of Gilbert, Henry, and Maxwell were seized upon by the inventors of the electric motor, the electric generator, the telegraph, telephone, and other devices. Not until the end of the nineteenth century were research methods applied to the exploitation of electrical phenomena, first by Thomas Edison who, in reality, was more of an inventor than a scientist, and later by many technologists in the laboratories of such industries as the General Electric Company and the predecessors of the American Telephone and Telegraph Company. Thus, it was in these fields—chemistry and electricity—that the merger of scientific inquiry and technology first occurred, that the power of the scientific methods was applied to solving useful problems, and that the great value of the thorough understanding of physical phenomena was demonstrated. In exploiting electrical phenomena technologists deal with fields and electrons and waves which can only be observed indirectly and understood through scientific research. It is not surprising, therefore, that scientifically based industry, like the electronics industry which depends very heavily upon basic research, should be the sponsor of much fundamental research.

Modern technology still requires invention. The vacuum tube, the transistor, memory devices for computers, and new materials tailored with specific properties are all inventions. But they are inventions made by men with special knowledge who have an understanding of a scientific field and who base their inventions on an intimate familiarity with that field, just as the inventor of old called upon his firsthand experience of the world he could see and feel to provide the working substance of his ingenuity.

This is the nature of modern, scientifically based technology. Clearly, the first requirement is the existence of a body of scientific knowledge. To use this knowledge as the basis for an invention in the solution of a specific problem, the technologist must have good understanding of the underlying science. Also, more likely than not, as he converges on the development of a specific device, the technologist will find that he is handicapped by the fact that the scientists who first explored the field that he is exploiting left vast areas of ignorance which must be filled before his task can be completed. These can only be filled by doing further fundamental research. Because the specific knowledge required to solve a problem is its goal, such work is often called "applied" or "directed research," though it is obvious that in another context it would be regarded as fundamental or basic research.

Who does this applied work? It depends upon the field. Generally, the exploitation of new areas of science, such as the recent efforts in solid state physics or nuclear physics, is initiated by scientists who are very well acquainted with the fundamentals of the field. These men usually create a reservoir of technological information and train the technologists, applied scientists, and engineers—men who ordinarily do not have as thorough a scientific background as the scientists who first exploited the field. One can see in this interplay a cause of much of the confusion that exists between scientific research and technology.

I have gone into this detail to show two things—the deep dependence of modern technology upon fundamental scientific knowledge and the interaction that must exist between scientist and technologist as new scientific information is employed for useful purposes. In planning scientific programs it is important to understand fully the essential difference between fundamental research done to achieve a deeper understanding of physical phenomena and technological efforts based upon such fundamental knowledge, but undertaken to meet a specific need or to solve a specific problem. Interesting and potentially useful basic research is often difficult to defend because it does not have a foreseeable application. Frequently, while Special Assistant to President Kennedy, I was confronted with an angry demand to explain how a particular piece of research could possibly be useful. This was hard to handle. If I merely said that basic research need not have a practical objective, I got into more trouble. I usually ended up giving a necessarily long discussion of the substance of the work, the unpredictability of scientific research, the long lead times involved in getting full understanding, and the educational value of the work, etc.

The research scientist is primarily motivated by an urge to explore and understand, but society supports fundamental research primarily, because experience has demonstrated how essential such work is for continuing progress in technology. Halt the flow of new research and the possible scope of technical developments will soon be limited and ultimately reduced to nothing. Incidentally, scientific knowledge need not be exploited immediately once it becomes available. It exists for all to use forever.

Thus, acquiring scientific knowledge is a form of capital investment. Unlike most other capital investments, it does not become obsolete nor can it be used up. Technological developments are also a form of capital investment, though somewhat less enduring. To be sure, a more efficient process will also yield its benefits endlessly, but usually technological developments tend to become obsolete as better methods, devices, and processes emerge.

From these facts one can derive the following guidelines for judging public investments in science and technology. Technological development should only be undertaken to fulfill specific needs, and only if the proposed new developments give promise of being economically justifiable as well as technically sound. Basic research should be judged primarily on its scientific merit and supported at a level which permits all meritorious work if available funds permit.

The past two decades have seen unprecedented expansion in research and development activities. United States' expenditures for research and development have doubled approximately every five years during this period and currently amount to about $19 billion per year, with nearly $15 billion of this amount provided by the federal government. Approximately 70 per cent of the federal funds are spent for defense and space activities, the remainder for a broad range of health, natural resource, basic research, transportation, and educational activities. Approximately 10 per cent of the total research and development funds have been used to support basic research activities.

As a consequence of this expansion there have been major advances in our understanding of the physical world and, at the same time, dramatic increases in the capabilities of technology. It would require much more time than I want to use to even list all of the important recent developments, but they include such scientific achievements as comprehensive understanding of atomic structure, a theory of the chemical bond, understanding of atomic nuclei, major advances in fundamental biological phenomena, particularly at a cellular and sub-cellular level, and the development of information theory and feedback and control theory. While in most fields understanding is by no means complete—in fact there is a long way to go in many of them—there is enough knowledge to permit almost miraculous technological achievements. Familiar ones, some already mentioned in connection with defense activities, include jet aircraft, television, atomic energy, radar, a vast array of new materials including plastics and metals, missiles and space craft, and the electronic computer.

Of all recent developments none is more revolutionary and far-reaching than the high-speed electronic computer. Created initially to facilitate routine numerical calculations, computers have now reached the point where they can perform logical operations as well. So powerful, versatile, and pervasive is the electronic computer that its impact has been likened to the effect of the introduction of machines during the industrial revolution. The computer has been used in three distinct ways, each of which has had an important influence. First, present high-speed machines are ten million times faster than a human being in performing many mathematical calculations and, consequently, make possible scientific and engineering studies and data-processing feats which were formerly beyond human imagination—let alone human achievement. Second, the computer has been put to work performing many menial or repetitive control functions in industrial and military operations. In some instances it merely substitutes for human operators; in others, it does things which humans could not possibly do, such as routing automatic long-distance telephone calls, ballistic missile guidance, rapid inventory control, etc. Third, the computer has introduced the possibility of modeling and thus studying complex dynamic systems. The development of this capability is one of the most significant intellectual advances of our time. Its utility extends from pragmatic design applications in the development of complicated integrated mechanisms such as an aircraft, a nuclear reactor, an inventory control system or even a more advanced computer, to sophisticated scientific investigation of economic behavior, language systems and learning.

Simulation, that is, experimenting with a computer model of an actual system, makes possible engineering designs which previously could only have been achieved by laborious and costly trial and error methods, at best, and in many instances probably not successfully completed at all. The stability problems encountered in high-speed flight are of such a nature, as are the neutron-shielding problems in atomic reactors.

I recently participated in a study which demonstrated the versatility of computers when used for simulation. Agricultural production in Pakistan is very low and is actually declining in some areas as a result of salt being deposited on the surface of the soil. This was caused, in large measure, by a rising water table, itself caused by many years of intensive irrigation. A group of scientists assembled for the purpose by Dr. Roger Ravelle, at the time Special Assistant for Science and Technology in the Interior Department, made computer models of the areas in question, studied pumping and other drainage methods for dealing with the salinity, and found a satisfactory technical solution to the problem. Fortunately they went further. They made a computer model of the Pakistan agricultural economy and found that even with a major effort to eliminate waterlogging the Pakistan food problem would not be solved. In fact, their studies showed that because of the growing population, the food situation would become increasingly desperate. Further simulation showed that a much better plan would be to use the available resources for a balanced program of soil improvement, agricultural development (including the introduction of fertilizers, weed killers, and pesticides), the development of suitable plant varieties, and proper marketing facilities. Paralleled with this, there needed to be a major effort to improve the human resources. This would require improvement in health and nutrition and a program of agricultural education. The studies showed further that with the available resources, an area of approximately one million acres was about an optimum size unit on which to apply this comprehensive approach. The studies also demonstrated that the agricultural yield of such an area would increase about 7 per cent per year for an indefinite period. Efforts are now going forward to implement the plan. I have elaborated on this study because it is one example of the exciting opportunities that simulation presents.

Continued progress in many technological and scientific fields is only possible because the high-speed computer has become available. A significant break-through in molecular biology, the determination of the structure of large molecules by X-ray examination, for which Watson and Crick recently received the Nobel Prize, was possible only because a very high-speed computer was available to process the data that were produced in the experiments. There is neither enough time to carry out such computations manually nor enough human computers. Furthermore, it would be prohibitively expensive to do such calculations even if sufficient numbers of trained calculators and mathematicians existed. To carry out one million simple mathematical operations, such as a single multiplication would cost $12,500 if done by a human being using a desk calculator, and only 23 cents on the most modern computer. It is interesting to note that as computer evolution continues, and faster, more versatile machines become available, the cost per single calculation drops so that with the succeeding years it becomes economically feasible to employ computers for more and more complicated purposes both in industry and in engineering and scientific activities. The development of this new tool to amplify the mental capabilities of humans will insure continued progress in many scientific and technological areas where complexity would otherwise almost completely limit further understanding.

As a result of our ability to control and exploit energy, manipulate materials, and communicate rapidly, fundamental economic realities have been completely altered. The scarcity of natural resources and the law of diminishing returns, as applied to the intensive use of human resources, were accepted as underlying assumptions of economic theory in the nineteenth century and led to a gloomy outlook insofar as prospects for the improvement in the lot of the masses were concerned. Today the picture is quite different. We can expect to fulfill our resource needs for the next century or more through the application of technology. We are living in a period of increasing productivity based upon more elaborate machines and higher levels of training of the labor force. Consequently, there is an increasing availability of goods and services for all levels of the society and a freeing of human resources for newly created tasks.

At this point I would like to call your attention to an important property of our technologically based economy that is vital to understand. The harder we work it; the stronger it grows. This means that there is a very real premium in forcing the system to function close to the limit, or as a power engineer would say, "at full load." Equally important is the need to provide sufficient support for science and technology to insure a continued improvement in productivity. Unfortunately, it is not possible to fix the appropriate level of research and development needed to insure the sustained growth of the economy. Three per cent of the gross national product in the United States is currently spent on research and development. While in Washington, I tried to find some way of determining the optimum rate of research and development investment for our country, but with little success. The only concept I have been able to apply is that of an upper limit. From the point of view of economic growth, increments in research and development investment should not exceed the increased productivity that they make possible. (Obviously, other motivations, such as defense might well lead a nation to exceed this level for a period, but this can only be done by diverting resources from other needs.) Also, it is undoubtedly wrong to judge health-related activities by purely economic measures. Economists tell us that one should increase research and development investments until a point is reached where the augmented output due to productivity increases just equals the incremental investments. To use this measure properly, it is necessary to actually predict the total future value of a given research and development expenditure. This is obviously not possible, so I will assume that the upper limit for research and development expenditures in a given year should be the entire increased output due to productivity increases in the same year. This is obviously very conservative, but at least it provides some guidance.

Unfortunately, not all of our current research and development activities add significantly to our scientific potential or industrial capabilities. If one eliminates those portions of the space and military research and development expenditures which do not significantly add to the fund of scientific knowledge or industrial know-how, approximately $6 billion of the federal research and development budget is left, which does make a substantial contribution to growth and productivity. To this should be added $4 billion per year invested in research and development by the private sector of the economy. Thus, I conclude that the nation is investing approximately $10 billion per year in research and development which leads to increased national product.

With a national output of $400 billion and a productivity increase of 4 per cent, relevant research and development expenditure should not exceed $16 billion per year. While there would be little point in going to this limit, at least not on a continuing basis, for none of the increased yield would be available for the general welfare, the foregoing figures show that it is economically feasible to consider a substantial increase in current research and development expenditures provided technically sound proposals and an available work force exist. The desirability of doing this is underscored by the dual fact that we will fall approximately $40 billion short of our potential output this year while 5 million Americans are unemployed. Since this condition is chronic in the United States, we can well afford to train more technicians, engineers, and scientists and employ them on activities which will contribute to future economic growth and the general welfare. Even if research and development activities were allowed to grow at the same rate that they did during the past decade, there would still be plenty of excess capacity to do the many other things we should undertake collectively to improve our country. We can build more schools, train more teachers and pay them better, provide more medical care and better housing and recreational facilities, and undertake more adequate conservation measures to preserve our resources. We could do all of these things and still have enough left over to provide considerably more aid than we do now to the developing nations. It would undoubtedly not be possible to match unused human and industrial capacity with needs so as to fully employ them, but we should have little difficulty in expending $20 billion per year of our surplus resources very effectively. What is more, developing the technique of making large-scale public investments without a military or security motivation would be a most invaluable end in itself.

What should we do? There are many very promising research and development activities which could contribute to our social and economic progress, but that have lost out in the competition for funds against more immediate problems and, as a consequence, have been only partially supported, or in some instances not undertaken at all. Some of the more vital activities and estimated annual costs are listed below:

Educational Research and Development	$100,000,000
Environmental Health Research & Technology	50,000,000
Civilian Technology	50,000,000
Natural Resources Development	100,000,000
Desalinization of Sea Water by Nuclear Energy	40,000,000
Oceanography	80,000,000
International Development (including medical education, nutrition, economics, industrial development, etc.)	100,000,000
Supersonic Aircraft Development	200,000,000

This is not an exhaustive list of research and technological areas worthy of support, but it does show the possibilities that exist for productive investments.

In addition to these specific areas, many fields of basic research have been working with insufficient funds and urgently need additional support for facilities, equipment, and staff. In this category are the fields of chemistry, the behavioral and social sciences, the communication sciences, low energy nuclear physics, mathematics, and the engineering sciences. I believe that it would take about $400 million in equipment and $150 million per year in research support to come near to realizing the potential in these fields—a 10 per cent increase. Not included are funds for academic uses of computers. Earlier I showed how the electronic computer was revolutionizing research and development by making possible research and engineering activities previously beyond human capabilities and by speeding up other tasks. Unfortunately, the cost of purchasing and operating adequate computational facilities exceed the financial capabilities of every academic institution. By a combination of federal research support, generous discounts by computer manufacturers, and use of local funds, many universities have managed to establish and operate extensive computer facilities in the past, but academic demands are increasing so rapidly that present means of financing these activities no longer meet the needs. I believe that the introduction of computer courses and the use of computational methods in teaching physical sciences, engineering, economics, and social sciences will so enhance the professional competence of the students involved that extraordinary measures should be taken to assist the colleges. Possibly computers should be included in the category of academic facilities that the federal government will help to finance under the College Aid Bill.

Taken altogether, the research and development activities that I am proposing would cost slightly more than $1 billion per year, less than 20 per cent of the space budget and only 5 per cent of the presently available unused resources in the economy, and far below the annual increment which would be economically sound.

I would like to turn my attention now to questions concerning technical manpower. Several times, estimates have been made of the number of scientists and engineers who will be needed by the major employers of technical manpower during the next decade. Unfortunately, examination of their prediction leads one to the conclusion that each industry has done no more than project a continuation of the historical growth pattern, so that the result is no more meaningful than an extrapolation based on the over-all manpower figures of the past. There are approximately 1.7 million professionally active scientists and engineers in the United States who are engaged in a broad range of activities—from sales engineering, manufacturing, and technical management to basic research. This number has been doubling every twelve years for several decades. Prior to World War II most of the individuals were employed by private institutions and paid with private funds. During World War II, and since, most of the growth in technical employment has been due to the expenditure of moneys by the federal government, principally for national security purposes but increasingly in other areas, such as space and health research as well. Will there be a further doubling of the technical labor force in the next twelve years? It is possible.

During the past few years there has been some concern that the federal government was starting new technical activities at a rate which exceeded the supply of technically trained specialists to man them. This fear, coupled with the discovery that the Soviet Union was training about twice as many people as the United States, led President Kennedy to ask his Science Advisory Committee to study the problem of the supply of scientists and engineers. A panel of the committee made an exhaustive study of the probable needs during the next decade and attempted to estimate the supply as well. I have already discussed the difficulty of accurately predicting the needs, but assuming that the past trend does continue, it is possible to judge whether or not there will be a sufficient number of graduates in science and engineering. The committee concluded that the enrollment trends are such that there will not be a numerical shortage of technical manpower, in fact the force will approximately double in the next twelve years.

Since completion of the study in 1963, there has actually been some unemployment among scientists and engineers due to changes in the defense program. While this condition is not general, it is sufficiently widespread to change concerns about shortages of trained manpower to a worry that we may be educating more students than will be needed in the late sixties. This could well be so unless new technical activities, such as those I listed earlier, are undertaken.

While the panel concluded that it could find no evidence of a serious manpower shortage, existing or potential, it did find a serious shortage of highly trained people. One of the panel's most significant conclusions was that our schools of higher education must make a determined effort to increase the number of students receiving advanced training, that is, masters and doctors degrees, particularly in the fields of mathematics, the physical sciences, and engineering. There appear to be shortages, too, in fields related to our rapidly increasing medical and public health endeavors.

In the fields of engineering, the physical sciences, and mathematics, we are currently preparing approximately 3,500 Ph.D.'s per year. The panel suggested that our goal should be to double this number by 1970, with proportionate growth in the number of professional degrees and masters degrees awarded. This goal was established primarily by estimates of our ability, rather than by an analysis of need. The panel believed that an even larger number is desirable, but would not be feasible.

To accomplish this goal will require a marked expansion in the present graduate schools and the creation of a number of new graduate centers, as well as, provision of training grants and fellowships.

Increased graduate training will require additional graduate research activities and consequently an increase in research funds. The program I outlined earlier provides for this.

IV

Science and Man's Place in the Universe

MICHAEL POLANYI

I WANT TO sketch for you a theory of knowledge which abandons the ideal of scientific detachment. I shall show that all knowledge is based on a measure of personal participation. This will affect our world view. The Aristotelian universe had a hierarchic structure, which was destroyed when science declared all manner of things to be determined by the same laws controlling their ultimate elements. I shall show that once the personal participation of the knower is firmly established, we can again acknowledge the existence of entities governed by principles which cannot be accounted for by the laws of inanimate nature. The distinction between things essentially higher and things essentially lower will be restored.

In such a cosmos man becomes, once more, the point towards which all nature converges and from which man aims beyond himself to his ideals. We shall start on this inquiry by considering the fact *that we can know more than we can tell.*

This fact seems obvious enough; but it is not easy to say exactly what it amounts to. We know a person's face, and can recognize him among a thousand, indeed among a million. Yet we usually cannot tell how we recognize a face we know. Most of such knowledge cannot be put into words. At the universities great efforts are spent in practical classes to teach students to identify cases of diseases and specimens of rocks, of plants, and of animals. All descriptive sciences study physiognomies which cannot be fully described in words, nor even by pictures. All this practical teaching must rely on the fact that the pupil's intelligence will enable him to recognize the relevant particulars of a physiognomy and their characteristic relationship in forming the physiognomy. Formal teaching fails to tell what the teacher knows, and the pupil must discover it for himself. Eventually, both will know something they cannot tell. I shall call this kind of knowledge *tacit.*

Gestalt psychology has demonstrated that we may know a physiognomy by integrating our awareness of its particulars without being able to identify these particulars. But some recent psychological experiments have perhaps shown more clearly the principal mechanism by which tacit knowledge is acquired. The general public has heard of these experiments as exposing the machinery of hidden persuasion. Actually, they are but elementary demonstrations of the faculty by which we apprehend the relation between two events, both of which we know, but only one of which we can tell.

Following the example set by Lazarus and McCleary in 1949, psychologists call the exercise of this faculty a process of "subception." The two authors in question presented a person with a number of nonsense syllables and after showing certain of the syllables, they administered an electric shock to the subject. Presently the person showed symptoms of anticipating the shock at the sight of "shock syllables"; yet on questioning he could not identify these syllables. He had come to know when to expect a shock, but he could not tell what made him expect it. He had acquired a knowledge similar to that which we have when we know a person by means we cannot tell.

Another variant of this phenomenon was demonstrated by Ericksen and Kuethe in 1958. They exposed a person to shock whenever he happened to utter associations to certain "shock words." Presently, the person learned to forestall the shock by avoiding the utterance of such associations, yet on questioning it appeared that he did not know that he was doing this. Here the subject invented a practical operation, but could not tell how he worked it. This kind of subception has the structure of a skill, for a skill combines elementary muscular acts which are not identifiable, according to relations that we cannot define.

These experiments show clearly what is meant by saying that one can know more than one can tell, because the experimental arrangement eliminates any suspicion of self-contradiction which may arise when anyone speaks of things he knows and cannot tell. This is avoided here by the division of roles between the subject and the observer. The experimenter observes that another person has a certain knowledge that he cannot tell, and so no one speaks of a knowledge he himself has and cannot tell.

We may carry forward then the following result. In both experiments that I have described subception was induced by electric shock. In the first series the subject was shocked after being shown certain nonsense syllables, and he learned to expect this event. In the second series he learned to suppress the uttering of certain associations which would evoke the shock. In both cases the knowledge of the shock-producing particulars remained tacit. The subject could not identify the particulars, yet he relied on his awareness of them for anticipating the electric shock.

Here we see the basic structure of tacit knowing. It always involves two kinds of things. We may call them the two terms of tacit knowing. In the subception experiments the shock-syllables and the shock-associations formed the first term, and the electric shock which followed them was the second term. After the subject had learned to connect these two terms, the sight of the shock-syllables evoked the expectation of a shock. In the second experiment this led to the utterance of the shock-associations being suppressed in order to avoid shock.

Why did this connection remain tacit? It would seem that this was due to the fact that the subject was riveting his attention on the electric shock. Thus he became aware of the shock-producing particulars only in their bearing on the electric shock. He learned to rely on his awareness of these particulars only for the purpose of attending to the electric shock.

This then is the *basic logical relation* between the first and the second term of a tacit knowledge. It combines two kinds of knowing. We know the electric shock forming the second term by attending to it, and hence its knowledge is specifiable. We know the shock-producing syllables only by relying on our awareness of them for attending to something else, namely the electric shock, and hence our knowledge of them remains tacit. We come to know these particulars, without becoming able to identify them. Such is the *functional relation* between the two terms of tacit knowing: *we know the first term only by relying on our awareness of it for attending to the second.*

In his book on freedom of the will, Austin Farrar has spoken of *disattending from* certain things for attending *to* others. I shall adopt a similar usage by saying that in an act of tacit knowing we *attend from* something for attending *to* something else; namely, *from* the first term *to* the second term of the tacit relation. In many ways the first term of this relation will prove to be nearer to us, the second farther away from us. Using the language of anatomy, I shall call the first term *proximal*, and the second term *distal*. *It is the proximal term, then, of which we have a knowledge that we may not be able to tell.*

In the case of a human physiognomy, I would say now that we rely on our awareness of its features for attending to the characteristic appearance of a face. We are attending *from* the features *to* the face, and thus may be unable to specify the features. And I would say, likewise, that we are relying on our awareness of a combination of muscular acts for attending to the performance of a skill. We are attending from these elementary movements *to* the achievement of their joint purpose, and hence are usually unable to specify these elementary acts.

We have here the *functional structure of tacit knowing.*

But we may ask: Does the *appearance* of the experimental setting—composed of the nonsense syllables and the electric shocks—not undergo some change when we learn to anticipate a shock at the sight of certain syllables? It does, in a subtle way. The expectation of a shock, which at first had been vague and uncertain, now becomes sharply fluctuating; it suddenly rises at certain moments and subsides between these moments. So we may say that, even though we do not learn to recognize the shock-syllables as distinct from other syllables, we do become aware of facing a shock-syllable in terms of the apprehension it evokes in us. In other words, we are aware of seeing these syllables in terms of that on which we are focusing our attention, which is the probability of an electric shock. Applying this to the case of a physiognomy, we may say that we are aware of its features in terms of the physiognomy to which we are attending. In the exercise of a skill we are aware of its several muscular moves in terms of the performance to which our attention is directed. We are aware in general of the proximal term of an act of tacit knowing in the appearance of its distal term; or, in other words, we are aware of that *from* which we are attending *to* another thing, in the appearance of *that thing.*

We may call this the *phenomenal structure of tacit knowing.*

There is a significance in the relation between the two terms of tacit knowing which combines its functional and phenomenal aspects. When the sight of certain syllables makes us expect an electric shock we may say that they *signify* the approach of a shock. This is their *meaning* to us. Therefore, when shock-syllables arouse an apprehension in us, without our being able to identify the syllables which arouse it, we know these syllables only in terms of their meaning. It is their meaning to which our attention is directed. It is in terms of their meaning that they enter into the appearance of that *to* which we are attending *from* them.

We may say, in this sense, that a characteristic physiognomy is the meaning of its features; which is, in fact, what we do say when a physiognomy expresses a particular mood. To identify a physiognomy amounts to relying on our awareness of its features for attending to their joint meaning. This may sound farfetched, because the meaning of the features is observed at the same spot where the features are situated, and hence it is difficult to separate mentally the features from their meaning. Yet, the fact remains, that the two are distinct, since we may know a physiognomy without being able to specify its particulars.

To see more clearly the separation of a meaning from that which has this meaning, we may watch the use of a probe to explore a cavern, or the way a blind man feels his way by tapping with a stick. For here the separation of the two is wide, and we can also observe here the process by which this separation gradually takes place. Anyone using a probe for the first time will feel its impact against his fingers and palm. But as we learn to use a probe, or to use a stick for feeling our way, our awareness of its impact on our hand is transformed into a sense of its point touching the objects we are exploring. Thus our interpretative effort transposes our meaningless feelings into meaningful ones and places these at some distance from the original feelings. We become aware then of the feelings in our hand in terms of their meaning located at the tips of the probe, or stick, to which we are attending. This is so also when we use a tool. We are attending to the meaning of its impact on our hands in terms of its effect on the things to which we are applying it.

We have here the *semantic aspect of tacit knowing.*

All meaning tends to be displaced *away from ourselves,* and that is my justification for using the terms "proximal" and "distal" to describe the first and second terms of tacit knowing.

From the three aspects of tacit knowing—the *functional,* the *phenomenal,* and the *semantic*—we can deduce a fourth aspect which tells us what tacit knowing is a knowledge of. This will represent its *ontological* aspect. Since tacit knowing establishes a meaningful relation between two terms, we may identify it with the *understanding* of the comprehensive entity which these two terms jointly constitute. Thus the proximal term represents the *particulars* of this entity, and we can say that we comprehend the entity by relying on our awareness of its particulars for attending to their joint meaning.

This analysis can be applied with interesting results to the case of visual perception. Physiologists have established that the way we see an object is determined by our awareness of certain efforts inside our body which we cannot feel in themselves. We are aware of these processes inside our body in terms of the position, size, shape, and motion of an object to which we are attending. In other words, we are attending *from* these internal processes *to* the qualities of things outside. These qualities are what those internal processes *mean* to us. The transposition of bodily experiences into the perception of things outside, may appear now as an instance of the transposition of meaning at a distance, which we have found present in varying degrees in all tacit knowing.

It may be thought that the feelings transposed by perception differ from those transposed by the use of tools or probes, by being hardly noticeable before their transposition. An answer to this—or at least part of an answer to it—is to be found in experiments extending subception to subliminal stimuli. Hefferline and collaborators have observed (1959) that when spontaneous muscular twitches, unfelt by the subject—but observable externally by a millionfold amplification of their action currents—were followed by the brief cessation of an unpleasant noise, the subject responded by increasing the frequency of the twitches, thus silencing the noise for a time. Tacit knowing is seen to operate here on an internal action that we do not feel in itself. We become aware of our operation of it only in the silencing of a noise. This demonstrates a process similar to that which I have assumed for explaining how we become aware of subliminal processes inside our body in the perception of objects outside.

There is a striking feature of the world as we see it to which we must turn next. It lies in the fact that each of us divides the entire universe into two parts, one being our body with which we identify ourselves and the other the totality of all other things which are not our body. The way the two parts are related in our knowledge of the world is illustrated by the process of perception. We make sense of the world by relying on our awareness of the impacts made by the world on our body and of the responses our body makes to these impacts. In other words, we know the world by attending to it *from* our body; and our body differs from all other objects in the world by being the only collection of things which we know almost exclusively by attending *from* them to other things, rather than by attending to them in themselves. This is what is meant by saying that we live in our body. For we live in our body by using it as an instrument for knowing things outside; our confidence in our knowledge of such things expresses our commitment to life in our body.

This logical relation, that links life in our body to our knowledge of things outside, can be generalized to further instances in which we rely on our awareness of certain things for attending to something else. We may recognize then that when we attend from a set of particulars to the whole which they form, we establish a logical relation between the particulars and the whole, similar to that which exists between our body and the things outside it. In other words, we recognize our faculty for using the parts of a whole for attending to the comprehensive entity which they form, as if they were our body when used for attending from it to objects outside. We may be prepared to consider then the act of comprehending a whole *as an interiorization of its parts,* by virtue of which *we come to dwell in these parts;* this indwelling being logically similar to the way we live in our body.

This expanded use of the term "indwelling" modifies the customary meaning of the word. It applies here in a logical sense as affirming that the parts of the external world, when interiorized, function in the same way as our body functions when we attend from it to things outside. In this sense we live in the particulars which we comprehend and similarly in the tools and probes which we use.

We live also in our intellectual tools and probes: to apply a theory for understanding nature is to interiorize it. We attend *from* the theory *to* things interpreted in its light. We do not understand a mathematical theory well, until we have learned to use it as a tool.

But the depth of indwelling varies greatly. The things we know and cannot tell about an old friend are deeper than what we know and cannot tell about a stone falling to earth. The knowledge of a falling stone comes close to the ideal of scientific detachment which demands that we explain all things in terms of mathematical physics. When applied to living beings, this would require that we explain them by the laws governing their ultimate particulars. The existence of higher levels knowable only by deeper indwelling would be excluded.

But look at biology. It studies the shapes of living things and the way they grow into these shapes from germ cells; it describes the organs of living things and explains the way they function; it explores the motoric and sensory functions of animals and their intelligent performances. All these are comprehensive entities. Morphology, physiology, animal psychology—they all deal with comprehensive entities. None of these entities can be mathematically defined and the only way to know them is to comprehend the coherence of their parts. This is how the existence of animals and plants was recognized long before zoology and botany; how the difference between life and death was recognized before physiology; and the differences between sentience and insentience, between intelligence and mindless stupidity, were known long before these matters were studied by science. What is more, this *still is* the way scientists themselves must recognize these things before they can undertake to study them.

But the false ideal of scientific detachment will not admit that scientists recognize living shapes and living functions by such indwelling and that they cannot recognize them otherwise. Biologists must insist that the only scientific way of representing living beings is in terms of the laws of physics and chemistry which govern their isolated elements.

This program combines a great absurdity with a more subtle error, both of which gain their convincing power by being identified with the triumphs of biology. Take first the absurdity. It is absurd to claim that the sentience of animals and the experience of consciousness in general can be accounted for by the laws of physics and chemistry. The laws of physics and chemistry do not ascribe consciousness to any process controlled by them; and if material processes could be found which—though otherwise controlled by the laws of physics and chemistry—were accompanied by consciousness, new laws would have to be discovered to account for this accompaniment.

Thus the explanation of all living processes by physics and chemistry collapses unless we deny altogether the existence of consciousness. It would seem impossible that neurophysiologists, let alone psychologists, should deny the existence of consciousness which is a major part of their subject matter. Can one study perception without referring to what people see? Or the localization of emotional centers in the brain without referring to what the subjects feel? Yet you find a distinguished neurophysiologist, like O. Hebb, urging scientists to assume that consciousness does not exist, even though such a hypothesis might eventually prove false. Nor is this an isolated instance. The psychiatrist L. S. Kubie spoke as follows on the same occasion.* He declared that a "working concept" of consciousness was indispensable to psychology, and went on to say:

> Sometimes we are explicit and frank about this. Sometimes we fool ourselves about it. Many workers have attempted to avoid using the word because of its traditional connotations, which have had a somewhat mystical, imponderable, non-scientific, philosophic and/or theological flavour.

Kubie's words show what is happening here. Scientists who urge us to assume that consciousness does not exist do not believe this themselves. It would be absurd to suppose that Hebb wants neurophysiologists to assume that all their subjects are unconscious. He merely wants them to describe their findings as if consciousness did not exist.

This is the program of behaviorism. It sets out for example to eliminate all references to the human mind, by substituting for the mind the sound of human speech telling about a state of mind. Such an inquiry refuses to ob-

* I am referring here to the contributions made by O. Hebb and L. S. Kubie to the Symposium on *Brain Mechanism and Consciousness* held in Quebec in August, 1953; see report ed. J. F. Delafrayne (Oxford: Blackwells, 1954).

serve that a man is in pain and can acknowledge only that he complains of pain. The fact that this view wipes out the purpose of medicine as the alleviator of human suffering is disregarded. Behaviorism could describe medicine only as a process for eliminating complaints of pain, even though complaints can be more effectively silenced without medicine. The very conception of compassion is denied, and torture is theoretically given free rein.

None of this is intended, or even remotely approved, by behaviorists who call in question the existence of consciousness. It is clear, therefore, that they do not mean what they say when urging us to doubt or disregard or at least avoid mentioning the existence of consciousness. This is what Kubie calls fooling ourselves. We take pride, as scientists, in professing something that laymen would find absurd. We feel ourselves then as successors to the Copernicans who forced laymen to see our earth, the very ground of fixity, hurtling around an immobile sun.

Such fooling of ourselves is widely admitted in biology. Everyone knows that you cannot inquire into the functions of living organisms without referring to the purpose served by them and by the organs and processes which perform these functions. Yet we must pretend that all such teleological explanations are merely provisional. The story goes round among biologists everywhere that teleology is a woman of easy virtue whom the biologist disowns in public, but lives with in private.

This particular mixture of pretence and self-deception is supported by a logical error which I mentioned earlier as the second, more subtle, absurdity in the professed program of biology. This error consists in asserting that biology explains living beings in terms of physics and chemistry, while the purpose which biology actually pursues, and by which it achieves its triumphs, consists in explaining the living beings in terms of a mechanism founded on the laws of physics and chemistry, but not determined by them.

The distinction is important and yet fairly simple. The laws of physics and chemistry determine the processes and structures of inanimate nature. We inquire into these processes and structures without asking what they are for, because we assume that they achieve nothing. Thus do we discover the laws of physics and chemistry. Take by contrast the study of physiology. It consists in observing the structures of living beings and the processes occurring in them, with a bearing on the question of what these things are for. For the task of physiology is to explain the achievements of living beings. It usually explains them in terms of a mechanism which functions in the way a machine works. Physiological mechanisms are based, like machines, on the laws of physics and chemistry, but they are not accounted for by these laws. They are determined by rules which, as in machines, may be called *operational principles.* Operational principles describe the parts composing a mechanism and define the way they function in making the mechanism work.

You may ask, how it is possible for a mechanism which obeys the laws of physics and chemistry to be determined also by another principle not accountable by physics and chemistry? The answer is that the physical sciences expressly leave open certain conditions of a system that are usually described as its *boundary conditions,* and that the operational principles of a mechanism take effect by controlling these boundary conditions. As a result, physical and chemical laws *are made to serve* the physiological mechanism of living beings.

This structure of a mechanism makes it liable to failure, for the physical process on which it relies will tend to escape its control. Machines can break down and all life is likewise in jeopardy of disease and death. This distinguishes mechanisms from the processes of inanimate nature which can never go wrong.

I shall return to this analysis of mechanisms when I come to speak of the evolutionary origin of man. Let me now turn to the question, whether my rejection of the professed ideals of science implies a rejection of its results. My answer is that the practice of science is usually sound, even when it is conducted in the name of false principles. For biologists to deny their use of teleological reasoning is quite harmless. It is even possible that some valuable research *must* be based on absurd assumptions. Think of the recent exploration of various parts of the brain by electrodes of microscopic size, which showed the nervous system operating as a machine. This splendid inquiry would be hampered by keeping in mind the fact that the assumption of the whole nervous system operating as an insentient automaton is nonsensical. Neurologists are right, therefore, in ignoring the absurdity of the idea underlying their work. The same may be true for many parts of science, but certainly not for all. In some fields, as we shall find in the theory of evolution, the ideal of strict detachment impairs the very conclusions of science. When applied to psychology this ideal tends to impoverish inquiry, and may obscure its results. A sociology which must ignore the power of our ideals in society must ignore what is most essential in society or deal with it only obliquely by a breach of its principles. Wherever the current scientific outlook bears directly on man and society it affects our world view, and in doing so, tends to render all things meaningless. Only the blessed inconsistency of its expositors prevents them from carrying out fully this logical consequence.

But we cannot continue to rely for our view of men and the universe on merciful right-wing deviations from the current scientific outlook. However widely the ideal of strict detachment may help to guide science to ever new discoveries, we must not allow it to deprive our image of man and the universe of any rational foundation. All men, scientists included, must seek and hold on to a reasonable view of the universe and of man's place in it. For this we must rely on a theory of knowledge which accepts indwelling as the proper way for discovering and possessing the knowledge of comprehensive entities.

In our search for a reasonable world view, we shall turn in the first place to common understanding. I have reminded you that science itself largely relies for its subject matter on a common knowledge of things. The conceptions of life and death, of plants and animals, of health and sickness, of youth and old age, of mind and body, of machines and technical processes, and of innumerable other equally important things are commonly known. All these conceptions apply to complex entities, the reality of which is called in question by a theory of knowledge, which claims that the whole universe should ultimately be represented in all its aspects by the physical laws governing the inanimate substrate of nature. The new theory of knowledge rejects this claim and restores our respect for the immense range of common knowledge acquired by indwelling. Starting from here, let me sketch out our cosmic perspective by exploring the wider implications of the fact that all knowledge is acquired and possessed by indwelling.

I have spoken of the way we identify a person's physiognomy by relying on our awareness of features which we cannot specify, and I have shown that, in a sense, this amounts to a dwelling in the features of a person for the purpose of comprehending their joint meaning. We can read in the features and behavior of a person also the presence of moods, the gleam of intelligence, and the signs of sanity and human responsibility. At a lower level, we comprehend by a similar mechanism the body of a person and understand the functions of his physiological mechanism. We have seen that even physical theories comprehend in this way the processes of inanimate nature. Such are the various levels of knowledge acquired and possessed by indwelling.

These levels form a hierarchy of comprehensive entities. Inanimate nature is comprehended by physical laws; the mechanism of physiology is built on these physical laws and enlists them in its service; next, the intelligent behavior of a person relies on the healthy functions of his body controlled by him and, finally, moral responsibility relies on the faculties of intelligence which it directs.

Remember how the operations of machines, and of mechanisms in general, rely on the laws of physics and chemistry, but cannot be accounted for by these laws. In a hierarchic sequence of comprehensive levels each higher level is related to the levels below it in the same way as the operations of a machine are related to the particulars obeying the laws of physics. You cannot explain the operations of an upper level in terms of the particulars on which its operations rely. Each higher level of integration represents, in this sense, a higher level of existence, not accountable by the levels below it.

Yet each higher level is known to us by relying on our awareness of the particulars on the level below it. *We know each level by interiorizing its particulars and mentally performing the integration which constitutes it.* This is how all knowledge is based on indwelling, and this is how the consecutive stages of indwelling form a continuous transition from the understanding of the inanimate to the understanding of man's moral responsibility. The sciences of the I-It relations thus pass imperceptibly into the sciences of the I-Thou relations. From the minimum of indwelling, exercised in a physical observation, we move without a break to the maximum of indwelling which is a total commitment.

The cosmic significance of this panorama is revealed when we look upon it as the stages of an evolution that has achieved the rise of man. It may seem obvious that the succession of changes, sustained over a thousand million years, which have transformed microscopic specks of protoplasm into the human race, has brought forth, in doing so, a higher and altogether novel kind of being; that this process ranks with the one by which a fertilized human germ develops into a mature infant and the infant grows into an adult—a process that obviously produces a higher form of being.

But scientists must deny this. Their theory of knowledge requires that all stages of life be accountable by the laws governing inanimate nature. Hence they cannot recognize the rise of higher levels of existence that are inexplicable in terms of the laws governing the lower levels. They must believe, or try to believe, that a germ cell can grow into a man writing sonnets, like Shakespeare, without developing any faculty that is not accountable by the laws of physics and chemistry.

In this light, evolution would represent the production and reproduction of a set of relatively stable atomic configurations. This is also all that a modern theory of evolution *can* account for. Darwinian selectionism has no place for an evolution continuously tending towards higher forms of life.

Writers have protested time and again against this denaturing of evolution. They have insisted on the fact that the central feature and problems of evolution lie in its sustained tendency to produce higher levels of existence. But these writers have always been silenced. Listen to the latest instance of this; how one of the most distinguished biologists of our time brushes aside Teilhard de Chardin's plea for the recognition of this central fact of evolution: ". . . the idea that evolution has a main track or privileged axis is unsupported by scientific evidence." [P. B. Medawar in *Mind* LXX (1961), p. 99.] This statement confirms my view that as long as science accepts the false ideal of strict detachment, it cannot but deny reality to the most significant features of the universe. The new theory of knowledge, combined with the logical distinction between levels of existence, should remedy this blindness by providing a conceptual framework which recognizes the emergence of ever higher levels of reality by evolution.

If we are to reconsider the process of organic evolution in this sense, we must first restore the problem from its misrepresentation by the current theory of evolution. I have said that for me the central problem of evolution lies in the rise of higher beings from lower ones and, above all, in the rise of man. A theory which recognizes only evolutionary changes due to the selective advantage of random mutations cannot acknowledge this problem. For the capacity to survive is no criterion of evolutionary achievement in my sense. There exist today animals and plants on every evolutionary level. The lower species have of course survived up to date much longer than the higher ones and so their proven powers of survival are the greater. But even if it could be shown that, for some reason, life at a higher level succeeds better in surviving than at a lower level, this would not explain how higher forms of life have come into existence, any more than the fact that living things emerging from the inanimate have continued to live explains the origin of life. The current theory of evolution could explain as easily—indeed *more* easily—the descent of the amoeba from man, as the actual rise of man from creatures like the amoeba. Hence it is not dealing with evolution at all.

The representation of evolution, as due to differential selective advantage, has been assisted by shifting attention from evolution to the origin of species. Preoccupation with the way novel populations come into existence had made us lose sight of the more fundamental question, how any single individual of a higher species ever came into existence. I shall bring this problem into focus by surveying the historical antecedents of any single individual of a higher form.

The origins of one man can be envisaged by tracing the man's family tree all the way back to the primeval specks of protoplasm in which his first origins lie. The history of this family tree includes everything that has contributed to the making of this human being. This segment of evolution is precisely on a par with the history of a fertilized egg developing into a mature man, or the history of a single plant growing from seed, which includes everything that caused that man, or that plant, to come into existence. Natural selection plays no part in the evolution of a single human being. We do not include in the mechanism of growth the possible adversities which did not befall it and hence *did not prevent it.* The same holds for the evolution of a single human being; nothing is gained in understanding this evolution by considering the adverse chances which might have prevented it.

The distinction between the origin of species and the evolutionary origin of a single individual is logically sharp. To represent changes in population as equivalent to the coming into existence of their members is like saying that you catch a tiger by catching two and letting one go. It might help to keep the two conceptions apart if we coin a new name for the evolution of a single individual and call it an *idiogenesis* as distinct from a *phylogenesis.*

Idiogenesis does not disregard the occurrence of accidental mutations which may prove adaptive. It merely assumes that these can be distinguished from changes of type achieving new levels of existence. Most palaeozoists would agree that, though this distinction is often difficult, it is none the less valid. And once this obvious distinction is allowed for, the autonomous evolutionary rise of an individual is as clearly manifest as the growth of an individual from a germ cell. Thus, in the very facts of science, the true meaning of which science at present ignores, common intelligence can recognize this evolutionary performance as surely as it has recognized, before science, so many other fundamental performances of life.

In this light, evolution shows man arisen by a creative power inherent in the universe. The immense ancestral travail that has borne man invests him with a cosmic responsibility. Michelangelo's image of Adam created at God's command becomes a more intelligent symbol of man's position in the world than a description of man as a chance collocation of atoms. The ideas of Samuel Butler, Lloyd Morgan, Bergson, and Whitehead which have pointed toward this conclusion are developed here on firmer grounds.

Each successive stage of emergence is more comprehensive, more meaningful than the last. Yet a higher faculty must always operate through the levels below it. It must enlist the laws controlling the lower levels in the service of its own higher principles, and the lower level which enables the higher one to operate through it will always limit the scope of these operations and menace them with failure. All our higher endeavors must work through our lower nature and are necessarily exposed thereby to corruption. You may recognize here the cosmic roots of tragedy and of man's fallen condition.

This relation of the higher to the lower applies once more when an upper level endeavors to reach beyond itself. If it be true that no higher level can be accounted for by the operation of a lower level, then no effort of ours can be truly creative in the sense of establishing a higher principle not intrinsic to our initial condition. Yet this is what all great art, great thought, and great action must aim at. And this is how these efforts have, in fact, built up the heritage in which our minds grow up and live.

Has man's intelligence then broken through the limits of his own powers? Yes and no. Inventive efforts can never fully account for their success; but the story of man's evolution testifies to a creative power that goes beyond that which we can account for in ourselves. This power can make us surpass ourselves. We exercise some of it in the simplest act of acquiring knowledge and holding it to be true. For, in doing so, we strive for intellectual control over things outside, in spite of our manifest incapacity to justify this hope. The greatest efforts of the human mind amount to no more than this. All such acts find their example in the Pauline scheme which imposes an obligation to strive for the impossible in the hope of achieving it by divine grace. I mention this here, for ever since I read Reinhold Niebuhr's *Lectures on the Nature and Destiny of Man,* some twenty years ago, my inquiries into the theory of knowledge were guided by the idea that the Pauline scheme represents best the mechanism of man's striving for ideals beyond his reach.

V

Presupposition in the Construction of Theories

GERALD HOLTON

1.

THE GENERAL TOPIC which has been assigned to me in this series, *Science and the Humanities,* is not merely a subject of academic discussion. It can be a cataract of certain disaster if we do not cling to some narrow but more solid ledge. Both the scientific and the humanistic studies have been subject to such rapid changes that one must select one's foothold with great care. To document how far and how fast things have been moving during the last few generations I hardly need do more than cite two instances.

In the introduction to the volume of essays in his collected works, one of the most eminent authors in this country in the last century wrote in his first paragraph:

> Our age is retrospective. It builds the sepulchres of the fathers. It writes biographies, histories, and criticism. The foregoing generations beheld God and nature face to face; we, through their eyes. Why should not we also enjoy an original relation to the universe? . . . Why should we grope among the dry bones of the past, or put the living generation into masquerade out of its faded wardrobe? The sun

> shines to-day also. There is more wool and flax in the fields. There are new lands, new men, new thoughts. Let us demand our own works and laws. . . .

The author was not some brash scientist but Ralph Waldo Emerson. And his call was all too fully answered. No scholar or man of letters today would begin the edition of his collected works with the sentence "Our age is retrospective."

The changes in science have, of course, been no less striking. If one had to select a single illustration, it might well be one that indicates how differently the pace and natural involvement of science and industry were viewed. H. V. Hayes, the head of the mechanical department of the American Bell Telephone Company (later the American Telephone and Telegraph Company) wrote in a report dated March 7, 1892:

> I have determined for the future to abandon [the theoretical work] of this department, devoting all of our attention to practical development of instruments and apparatus. I think the theoretical can be accomplished quite as well and more economically by collaboration with the students of the [Massachusetts] Institute of Technology and possibly of Harvard College.[1]

And fifteen years later, in 1907, he wrote:

> Every effort in the Department is being executed toward perfecting the engineering methods; no one is employed who as an inventor is capable of originating new apparatus of a novel design . . . . I have felt that this policy was the wisest one . . . . The very fact that

[1] Federal Communications Commission Investigation, Pursuant to Public Resolution #8, 74th Congress, Docket #1, Bell System, Exhibit # 1951-A, vol. 1, p. 47.

any great invention at the present must in all probability come from some man of unusual scientific attainments would render a laboratory under the guidance of such men a most expensive and probably unproductive undertaking.[2]

2.

My thesis in this lecture will be that the dichotomy between scientific and humanistic scholarship, which is evident, undoubted, and real at many levels, becomes far less impressive if one looks carefully at the decision-making process in the construction of scientific theories. This is particularly true at the place where explicit and implicit decisions are most telling—namely in the formation, testing, and acceptance or rejection of hypotheses.

Current opinion on the way scientific theories are constructed is by no means unanimous. We may, nevertheless, take as an example the account given not too long ago by the physicist Friedrich Dessauer as a quite typical contemporary presentation of the so-called hypothetico-deductive, or inductive, method of science. His scheme[3] reflects both general professional and popular understanding.

There are, he reports, five steps. (1) Tentatively, propose as a hypothesis a provisional statement obtained by induction from experience and previously established knowledge of the field. An example, drawn from experimental work in physics, might be this: the observed large loss of sound energy when ultrasonic waves pass through a liquid such as water is possibly due to a structural rearrangement of the molecules as the sound wave passes

[2] *Ibid.*, p. 105. I wish to thank Mr. S. Goldberg for bringing both passages to my attention.

[3] *Eranos Jahrbuch,* XIV, 282 ff., 1946.

by them. (2) Now, refine and structure the hypothesis—for example, by making a mathematical or physical analogon showing the way sound energy may be absorbed by clumps of molecules. (3) Next, draw logical conclusions or predictions from the structured hypothesis which have promise of experimental check—for example, if more and more pressure is applied to the sample of water, it should be more and more difficult for the associated molecular groups to absorb sound so strikingly. (4) Then check the predicted consequences (deduced from the analogon) against experience, by free observation or experimental arrangement. (5) If the deduced consequences are found to correspond to the "observed facts" within expected limits—and not only these consequences, but all different ones that can be drawn (e.g., behavior at constant pressure but changing temperature, or similar effects in other liquids)—then, a warrant is available for the decision that "the result obtained is postulated as universally valid" (p. 298). Thus, the hypothesis, or initial statement, is found to be scientifically "established."

But, popular opinion continues, until the facts support such a position any hypothetical statement is to be held scrupulously with open-minded skepticism. The scientist, Dessauer reports, "does not take a dogmatic view of his assumption, he makes no claim for it, he does not herald it abroad, but keeps the question open and submits his opinion to the decision of nature itself, prepared to accept this decision without reserve" (p. 296). This, he concludes, is "the inductive method, the fundamental method of the entire modern era, the source of all our knowledge of nature and power over nature" (p. 301).

We note that this account fits in well with a widespread characterization of a supposed main difference between scientists and humanists: the former, it is often said, do not pre-empt fundamental decisions on esthetic or intuitive grounds; they do not make a priori commitments, and only let themselves be guided by the facts and the careful process of induction. It is, therefore, not surprising that in this, as in most such discussions, nothing was said about the *source* of the original induction, or about the criteria of *preselection* which are inevitably at work in scientific decisions. Attention to these would seem to be as unimportant or fruitless as a discussion, say, of the "reality" of the final result. To paraphrase Newton's disclaimer in the General Scholium of the *Principia,* to us it is enough that, for example, sound absorption in water does really exist and act according to the laws which we have explained, and abundantly serves to account for the pressure and temperature dependence of absorption in a great variety of liquids, even of the sea.

This account of scientific procedure is not wrong; it has its use, for example, in characterizing broadly certain features of science as a public institution. But if we try to understand the actions and decisions of an actual contributor to science, the categories and steps listed above are deficient because they leave out an essential point: to a smaller or larger degree, the process of building up an actual scientific theory requires explicit or implicit decisions, such as the adoption of certain hypotheses and criteria of preselection that are not at all scientifically "valid" in the sense previously given and usually accepted.[4]

[4] Some of the arguments presented here were first considered in my George Sarton Memorial Lecture, presented on December 28, 1962 before the meeting of the American Association for the Advancement of Science; others in a lecture of November 11, 1963 to the American Philosophical Association. See also my article "Über die Hypothesen, welche der Naturwissenschaft zu Grunde liegen," *Eranos Jahrbuch,* XXXI (Zurich: Rhein-Verlag, 1963), pp. 351–425.

3.

To illustrate this point as concretely as possible, let us look at a case for which it has long been thought the last word had been said. As is well known, Book III of Newton's *Principia,* which was supposed to use the principles and mathematical apparatus developed in Books I and II to "demonstrate the frame of the System of the World," opens with a section that is as short as it is initially surprising: the four rules of reasoning in philosophy, the Regulae Philosophandi. At any rate, they appear so in the third edition, of 1726, known to us usually through Motte's translation of 1729. These are, of course, well-known rules, and I need remind you of them only briefly. They can be paraphrased as follows:

I. Nature is essentially simple; therefore, we should not introduce more hypotheses than are sufficient and necessary for the explanation of observed facts. This is a hypothesis, or rule, of simplicity and *verae causae.*

II. Hence, as far as possible, similar effects must be assigned to the same cause. This is a principle of uniformity of nature.

III. Properties common to all those bodies within reach of our experiments are to be assumed (even if only tentatively) as pertaining to all bodies in general. This is a reformulation of the first two hypotheses, and is needed for forming universals.

IV. Propositions in science obtained by wide induction are to be regarded as exactly or approximately true until phenomena or experiments show that they may be corrected or are liable to exceptions. This is a principle that propositions induced on the basis of experiment should not be confuted merely by proposing contrary hypotheses.

It has been justly said that these epistemological rules are by no means a "model of logical coherence."[5] They grew in a complex way, starting from only two rules (I and II) in the first edition of the *Principia* (1687) where they were still called Hypotheses I and II. As Newton, with growing dislike for controversy, came to make the corrections for the third edition, he added the polemical rule IV which is a counterattack on the hypotheses-laden missiles from the Cartesians and Leibnizians.

But it turns out that Newton at one time was on the verge of going further. It was discovered only recently in a study of Newton's manuscripts by Koyré[6] that Newton had written a lengthy *Fifth Rule;* and then had suppressed it. The significant parts of it for our purpose are the first and last sentences of this rule, and the likely reasons why it had to be suppressed.

"Rule V. Whatever is not derived from things themselves, whether by the external senses or by internal cogitation, is to be taken for hypotheses . . . . And what neither can be demonstrated from the phenomena nor follows from them by argument based on induction, I hold as hypotheses."

To us, even as to Newton's contemporaries, disciples, and defenders, the sense in which Newton uses here the word "hypothesis" in the suppressed rule is clearly pejorative. It was after all Newton himself who, in 1704, had written as the first sentence of the *Opticks,* "My design in this Book is not to explain the Properties of Light by

[5] Cf. Alexandre Koyré, "Etudes Newtoniennes I.—Les regulae philosophandi," *Archives Internationales d'Histoire des Sciences,* **13** (1960), p. 6. This article and Koyré's "L'Hypothèse et l'experience chez Newton," *Bull. Soc. Française de Philos.,* **50,** 2 (1956), pp. 60–97, are perhaps the best introduction to the large literature on this subject.

[6] Koyré, "Etudes Newtoniennes I," *loc. cit.*

Hypotheses, but to propose them, and prove them by Reason and Experiment." And in this and other ways, he had begun to sound the declaration "hypotheses non fingo" in the second, 1713 edition of the *Principia.* We are apt to remember this slogan rather than the fact that in Newton's work from beginning to end, and even in the last edition of the *Principia* itself, one can readily find explicit hypotheses as well as disguised ones. And we are apt to overlook that rules against hypotheses are themselves methodological hypotheses of considerable complexity.

But, then, is it not strange that Newton after all *did* suppress this fifth rule which the Newtonians after him, his modern, empiricist disciples, from Cotes to Dessauer, would accept readily? To understand why Newton may have done this is of importance if we want to understand the cost of having so long been the philosophical heirs of the victorious side in that seventeenth-century quarrel concerning what science should be like.

The answer has, I think, several elements, but one is surely an ancient one: that disciples are usually eager to improve on the master, and that the leader of a movement sometimes discovers he cannot or does not wish to go quite as fast to the Promised Land as those around him. (Thus, it was not Cortes but the men he had left in charge of Mexico who, as soon as his back was turned, tried to press the victory too fast to a conclusion, and began to slaughter the Aztecs, with disastrous consequences.)

Here it is significant that Newton had only said, in one draft of his *General Scholium:* "I avoid hypotheses;" and in the final version, "I do not feign hypotheses," i.e., I make no false hypotheses. But his spokesman and friend, Samuel Clarke, translated him to read: "And hypotheses I *make* not;" and Andrew Motte rendered it as the famous "I frame no hypotheses." In this, as in several other places, Newton's protagonists went much further than he did, and seemed to ask for a Baconian sense of certainty in science which Newton knew did not exist.

Newton had indeed exposed and rejected certain hypotheses as detrimental; he knew how to tolerate others as being at least harmless; and he, like everyone else, knew how to put to use those that are verifiable or falsifiable. But the fact is that Newton also found one class of hypotheses to be impossible to avoid in his pursuit of natural philosophy—a class that shared with Cartesian hypotheses the characteristic of neither being demonstrable from the phenomena nor following from them by an argument based on induction, to use the language in Newton's suppressed fifth rule itself. The existence, nay, the necessity, at certain stages, of entertaining such unverifiable and unfalsifiable, and yet not quite arbitrary, hypotheses—that is an embarrassing conception which did not and does not fit into a purely positivistically oriented philosophy of science. For the decision whether to entertain such hypotheses is coupled neither to observable facts nor to logical argument.

In Newton's case, two obvious examples of his use of this class of hypotheses—to which I refer as "thematic" propositions or thematic hypotheses, for reasons to be discussed later—appear in his theory of matter and his theory of gravitation. On the latter, A. R. Hall and M. B. Hall, in their recent book *Selections from the Unpublished Scientific Papers of Sir Isaac Newton* (Cambridge, 1961), have printed the first manuscript draft of the *General Scholium* (written in January, 1712–13) in which Newton very plainly confesses his inability to couple the hypothesis of gravitational forces with observed phenomena: "I have not yet disclosed the cause of gravity, nor have I undertaken to explain it, since I could not understand it from the phenomena. For it does not arise from the centrifugal force of any vortex, since it does not tend to the axis of a vortex but to the center of a planet" (p. 352). And speaking of Newton's inability to arrive at the cause of gravity from phenomena, the Halls add: "In one obvious sense, this is true, and in that sense it knocks the bottom out of the aethereal hypothesis. In another sense it is false: Newton knew that God was the cause of gravity, as he was the cause of all natural forces. . . ."[7]

Exactly so—for this indeed was Newton's central presupposition in the theory of gravitation. The Halls continue, "That this statement could be both true and false was Newton's dilemma: In spite of his confident expectation, physics and metaphysics (or rather theology) did not smoothly combine. In the end, mechanism and Newton's conception of God could not be reconciled . . . . Forced to choose, Newton preferred God to Leibniz."

That Newton could not bring himself to announce this hypothesis in the *Principia* is not strange since the grounds of the hypothesis are so foreign to the avowed purpose of this book on the *Mathematical Principles of Natural Philosophy*. And also, a thematic hypothesis becomes the more persuasive the longer the period of unsuccessfully trying to use other hypotheses, namely, those that *are* coupled to phenomena. The thematic hypothesis is often an impotency proposition, in the sense that the search for alternatives has proved to be vain. The point when one is

[7] Or perhaps more precisely, in Newton's thought, as A. Koyré said, the cause of gravity is the action of the "Spirit of God." *From the Closed World to the Infinite Universe* (New York: Harper and Brothers, 1958), p. 234.

forced to rely on thematic hypotheses is exactly when one has to say, with Newton: "I could not understand it from the phenomena."

So when we approach the physics of a man like Newton, and even when we try to interpret his epistemological position, we must look beyond the explicit and obvious component of it, the basically operationist and relativistic physics of observable events. What made his work meaningful to Newton was surely that in his physics he was concerned with a God-penetrated, real world: God himself is standing behind the scenes, like a marionette player, moving the unseen strings of the puppets that merely act out the thoughts in His great sensorium. And this is a proposition which Newton tried to avoid having to state openly, where his friends and enemies would see it, though this reluctance accounts for some of the strange tension which pervades the *Principia* and his other writing. Reading Newton, one is struck by the fact that below the surface the major problems which haunted him were very closely related. They are: (a) the cause of gravity, whose existence only he had "established from phenomena"; (b) the existence of other forces, e.g., short-range forces to explain cohesion, chemical phenomena, etc.; (c) the nature of space and time, what he called the "sensory" of God; (d) and last, but not least, the proofs for the existence of the Deity (namely, by showing that there can be no other final causes for demonstrated forces and motions than the Deity—that, therefore, the Deity not only has properties, but also "dominion").

In Newton's physics, the hypothesized "sensory" of God is the cut-off point beyond which it was unnecessary and inappropriate to ask further questions. And this is an important function of a thematic hypothesis, which by its very nature is not subject to verification or falsification. For unlike the usual class of hypothesis—which, to use Aristotle's formula, is a statement that may be "believed by the learner" but ultimately is "a matter of proof"—the thematic hypothesis is precisely built as a bridge over the gap of ignorance. Thus, as scientists, we cannot and need not ask *why it is* that we believe, with Descartes, in an "inescapably believable" proposition; or why it is that we can perceive correspondences between certain observations and the predictions that follow from a model; or nowadays, for that matter, with Niels Bohr, why we can "build up an understanding of the regularities of nature upon the consideration of pure number."

4.

We have indeed left the recipe for a step-by-step construction of scientific theory far behind. Let us now turn from the specific example and attempt to discern in a schematic way what the analysis of scientific theories in terms of themata adds to the more conventional kind of analysis.

Regardless of what scientific statements they believe to be "meaningless," all philosophies of science agree that two types of proposition are *not* meaningless, namely statements concerning empirical matters of "fact" (which ultimately boil down to meter readings), and statements concerning the calculus of logic and mathematics (which ultimately boil down to tautologies).

There are clearly difficulties here that we might well discuss. For example, the empirical matters of fact of modern science are not simply "observed," but are nowadays more and more obtainable only by way of a detour of technology (to use a term of W. Heisenberg) and a detour of theory. But in the main we can distinguish between these two types of "meaningful" statements quite well. Let us call them respectively *empirical* and *analytical* statements, and think of them as if they were arrayed respectively on orthogonal x- and y-axes, thereby representing these two "dimensions" of usual scientific discourse by a frank analogy.

Now we can use the x-y plane to analyze the concepts of science (such as force), and the propositions of science, e.g., a hypothesis (such as "X-rays are made of high energy photons") or a general scientific law (such as the law of universal gravitation). The *concepts* are analogous to points in the x-y plane, having x and y co-ordinates. The *propositions* are analogous to line elements in the same plane, having projected components along x and y axes.

To illustrate, consider a concept such as force. It has empirical, x-dimension meaning because forces can be qualitatively discovered and, indeed, quantitatively measured, by, say, the observable deflection of solid bodies. And it has analytical, y-dimension meaning because forces obey the mathematics of vector calculus (e.g., the parallelogram law of composition of forces), rather than, for example, the mathematics of scalar quantities.

Now consider a proposition (a hypothesis, or a law): the law of universal gravitation has an empirical dimension or x component—for example, the observation in the Cavendish experiment where massive objects are "seen" to "attract" and where this mutual effect is measured. And the law of universal gravitation has an analytical or y component, the vector-calculus rules for the manipulation of forces in Euclidean space.

An interpolation is here in order, to avoid the impression that there is some absolute meaning intended for the x- or y-components. Indeed, it is preferable to use the term "heuristic-analytic" for the y dimension, on grounds which I can at least indicate by noting that there exist in principle infinitely many possible logical and mathematical systems, including mutually contradictory ones, from which we choose those that suit our purposes. On the x axis we do not appear to have this degree of freedom to make "arbitrary" decisions on heuristic grounds. At least at first glance, we seem constrained to deal with the phenomena of our natural world as they present themselves to us, rather than with many mutually contradictory worlds of phenomena from which we might be free to select those to which we wish to pay attention. However, one can at least imagine worlds that are quite differently constructed, where on the one hand an infinitely large pool of phenomena contains "contradictory" sets (i.e., stones that sometimes fall and sometimes rise, in some random sequence), but where on the other hand our logical and mathematical tools are severely restricted—say, only to Aristotelian syllogisms and elementary arithmetic. Then we would be forced to select from all possible observables those which can be represented and discussed in terms of scalar quantities, and we would have to exclude forces, acceleration, momenta, etc. In that case, the x-dimension could be named the dimension of heuristic-empirical statements.

Now, to some extent we *are* in this situation even now in our "real" world. We get a hint of it when we think of the great number of phenomena that are thought to be important today, but that were unknown yesterday;[8] or if

[8] Or, conversely, observables that were important may become unimportant, as in chemistry, where the fundamental attention to the appearance and color of the flame in violent chemical reactions was given up with the phlogiston theory.

we think of the continual change in the allegory (for example, the allegory of motion itself), from the Aristotelian conception which equated motion and change of any kind, to the modern, much attenuated idea of motion as the rate of change of distance or displacement with respect to time, or quantifiable local motion.

We realize the same point also when we think of all the "phenomena" which at any time are simply not admitted into science—for example, heat and sound in Galileo's physics, or most types of single-event occurrences that do not promise experimental control or repetition in modern physical science. In short, we are always surrounded on all sides by far more "phenomena" than we can use, and which we decide—and must decide—to discard at any particular stage of science.

The choice of allowable *analytical* systems is in principle also very large. Thus, any point, on any object, could for purposes of kinematical description be regarded as the center of the world. But the choice, in practice, is quite restricted. Indeed, the reason why science, to the late nineteenth century, was so sure of the uniqueness of the given world is to be sought in the fact that the analytical systems then available were so simple and had so long remained without fundamental qualitative changes and alternatives. Thus Newton could say in the preface of the *Principia* that geometry itself is "founded in mechanical practice and is nothing but that part of universal mechanics which accurately proposes and demonstrates the art of measuring." This impression helped to reinforce the feeling that the world, found and analyzed by science in terms of then current x and y components, existed in a unique, a priori way. In mathematics one calls such a situation, where the potential plurality of solution shrinks to one or a very few, a "degenerate" case. It is only after the discovery of non-Euclidean mathematics that one begins to see the essential arbitrariness of the y-dimension elements in which our scientific statements are couched, and that one becomes open to the suggestion that there is also an arbitrariness in the decisions about what x-dimension elements to select. This recognition is perhaps at the heart of the current agnosticism concerning the old question as to the "reality" of the world described in the x-y plane.

But whether they are arbitrary or not, the x-y axes have, since the seventeenth and eighteenth centuries, more and more defined the total allowable content of science, and even of sound scholarship generally. Hume, in a famous passage, expressed eloquently that only what can be resolved along x- and y-axes is worthy of discussion:

> If we take in our hands any volume; of divinity, or school metaphysics, for instance: Let us ask, Does it contain any abstract reasoning concerning quantity or number? No. Does it contain any experimental reasoning concerning matter of fact or criteria? No. Commit it then to the flames. For it can contain nothing but sophistry and illusion.

If we now leave the x-y, or *contingent*,[9] plane, we are going off into an undeniably dangerous direction. For it must be confessed at once that the tough-minded thinkers who attempt to live entirely in the x-y plane are more often than not quite justified in their doubts about the

[9] One may call the x-y plane the *contingent* plane because the meaning of concepts and statements in it are contingent on their having both empirical and analytical relevance. Contingency analysis is thus the study of the relevance of concepts and propositions in x- and y-dimensions. It is a term equivalent to operational analysis in its wider sense.

claims of the more tender-minded people (to use a characterization made by William James). The region below or above this plane, if it exists at all, might well be a muddy or maudlin realm, even if the names of those who have sometimes gone in this direction are distinguished. As Dijksterhuis has said:

> Intuitive apprehension of the inner workings of nature, though fascinating indeed, tend to be unfruitful. Whether they actually contain a germ of truth can only be found out by empirical verification; imagination, which constitutes an indispensable element of science, can never even so be viewed without suspicion.[10]

And yet, the need for going beyond the x-y plane in understanding science and, indeed, in doing science, has been consistently voiced since long before Copernicus, who said that the ultimate restriction on the choice of scientific hypotheses is not only that they must agree with observation but also "that they must be consistent with certain preconceptions called 'axioms of physics,' such that every celestial motion is circular, every celestial motion is uniform, and so forth."[11] And if we look carefully, we can find even among the most hard-headed modern philosophers and scientists a tendency to admit the necessity and existence of a non-contingent dimension in scientific work. Thus Bertrand Russell[12] speaks of cases "where the premises of sciences turn out to be a set of pre-suppositions

[10] *The Mechanization of the World Picture* (Oxford: Clarendon Press, 1961), p. 304.

[11] Quoted from E. Rosen, *Three Copernican Treatises* (New York: Dover Publications, Inc., 1959), p. 29.

[12] *Human Knowledge* (London: G. Allen & Unwin, 1948), Part 6, Ch. 2.

tions neither empirical nor logically necessary;" and in a remarkable passage, Karl R. Popper[13] confesses very plainly to the impossibility of making a science out of only strictly verifiable and justifiable elements:

> Science is not a system of certain, or well-established, statements; nor is it a system which steadily advances towards a state of finality . . . . *We do not know: we can only guess.* And our guesses are guided by the unscientific . . . faith in laws, in regularities which we can uncover—discover. Like Bacon, we might describe our own contemporary science—"the method of reasoning which men now ordinarily apply to nature"—as consisting of "anticipations, rash and premature" and as "prejudices."

One could cite and analyze similar opinions by a number of other scientists and philosophers. In general, however, there has been no systematic development of the point. But it is exactly here that we should discern the existence of a door at the end of the corridor through which the philosophy of science has recently been traveling. To supplement contingency analysis, I suggest a discipline that may be called thematic analysis of science (by analogy with thematic analyses that have for so long been used to great advantage in scholarship outside science). In addition to the empirical or phenomenic (x) dimension and the heuristic-analytic (y) dimension, we can define a third, or z-axis. This third dimension is the dimension of fundamental presuppositions, notions, terms, methodological judgments and decisions—in short, of themata or themes—which are themselves neither directly evolved from, nor

[13] *The Logic of Scientific Discovery* (New York: Basic Books, Inc., 1959), pp. 279–80.

resolvable into, objective observation on the one hand or logical, mathematical, and other formal analytic ratiocination on the other hand. With the addition of the thematic dimension, we generalize the plane in which concepts and statements were previously analyzed. It is now a three-dimensional "space"—using the terms always in full awareness of the limits of analogy—which may be called *proposition space.* A concept (such as force), or a proposition such as the law of universal gravitation, is to be considered, respectively, as a point or as a configuration (line) in this threefold space. Its resolution and projection is in principle possible on each of the three axes.

To illustrate: the phenomenic and analytic-heuristic components of the physical concept force (its projections in the x-y plane) have been mentioned. We now look at the thematic component, and see that throughout history there has existed in science a "principle of potency." It is not difficult to trace this from Aristotle's ἐνέργεια, through the neo-Platonic *anima motrix,* and the active *vis* that still is to be found in Newton's *Principia,* to the mid-nineteenth century when "Kraft" is still used in the sense of energy (Mayer, Helmholtz). In view of the obstinate preoccupation of the human mind with the theme of the potent, active —I would almost say masculine—principle, before and quite apart from any science of dynamics (and also with its opposite, the passive, persisting principle on which it acts), it is difficult to imagine any science in which there would not exist a conception of force (and of its opposite, inertia).

It would also be difficult to understand certain conflicts. Scholastic physics defined "force" by a projection in the phenomenic dimension that concentrated on the observation of continuing terrestrial motions against a constantly acting obstacle; Galilean-Newtonian physics defined "force" quite differently, namely, by a projection in the phenomenic dimension that concentrated on a thought experiment such as that of an object being accelerated on a friction-free horizontal plane. The projections above the analytic dimension differed also in the two forms of physics (i.e., attention to magnitudes versus vector properties of forces). On these two axes, the concepts of force are entirely different. But the reason why natural philosophers in the two camps in the early seventeenth century thought they were speaking about the same thing, nevertheless, is that they shared the need or desire to incorporate into their physics the same thematic conception of *anima,* or *vis,* or *Kraft*—in short, of force.

A second example of thematic analysis might be the way one would consider not a concept but a general scientific proposition. Consider the clear thematic element in the powerful laws of conservation of physics, for example the law of conservation of momentum, as formulated for the first time in useful form by Descartes. In Descartes' physics, as Dijksterhuis wrote:[14]

> All changes taking place in nature consist in motions of . . . three kinds of particles. The primary cause of these motions resides in God's *concursus ordinarius,* the continuous act of conservation. He so directs the motion that the total *quantitas motus* (momentum), i.e., the sum of all the products of mass and velocity, remain constant.

"This relation Σ mv = const., constitutes the supreme natural law. . . ."[15] This law, Descartes shows, springs from the invariability of God, in virtue of which, now that

[14] Dijksterhuis, *The Mechanization of the World Picture,* p. 410.
[15] Descartes, *Principia Philosophiae* II, c. 36; *Oeuvres* VIII, 62–65.

He has wished the world to be in motion, the variation must be as invariable as possible.

Since then, we have learned to change the *analytic* content of the conservation law—again, from a scalar to a more complex calculus—and we have extended the phenomenic applicability of this law from impact between palpable bodies to other events (e.g., scattering of photons). But we have always been trying to cling to this and to other conservation laws, even at a time when the observations seem to make it very difficult to do so. Poincaré clearly saw this role of themata in a passage in *Science and Hypothesis:*[16] "The principle of the conservation of energy simply signifies that there is a *something* which remains constant. Whatever fresh notions of the world may be given us by future experiments, we are certain beforehand that there is something which remains constant, and which may be called *energy*"—(to which we now add: even when we used to call it only mass). The *thema* of conservation has remained a guide, even when the language has had to change. We now do not say the law springs from the "invariability of God;" but with that curious mixture of arrogance and humility which scientists have learned to put in place of theological terminology, we say instead that the law of conservation is the physical expression of the elements of constancy by which Nature makes herself understood by us.

The strong hold that certain themes have on the mind of the scientist helps to explain his commitment to some point of view that may in fact run exactly counter to all accepted doctrine and to the clear evidence of the senses. Of this no one has spoken more eloquently and memorably

[16] (1952 reprint, New York: Dover Publications, Inc.), p. 166. Emphasis in original.

than Galileo when he commented on the fact that to accept the idea of a moving earth one must overcome the strong impression that one can "see" that the sun is really moving:

> Nor can I sufficiently admire the eminence of those men's intelligence [Galileo's Salviati says in the Third Day of the *Dialogue Concerning the Two Principal Systems*], who have received and held it [the Copernican system] to be true, and with the sprightliness of their judgments have done such violence to their own senses that they have been able to prefer that which their reason dictated to them to that which sensible experience represented most manifestly to the contrary . . . . I cannot find any bounds for my admiration how reason was able, in Aristarchus and Copernicus, to commit such rape upon their senses as, in spite of them, to make itself master of their belief.

Among the themata which permeate Galileo's work and which helped reason to "commit such rape upon their senses," we can readily discern the then widely current thema of the once-given real world which God supervises from the center of His temple; the thema of mathematical nature; and the thema that the behavior of things is the consequence of their geometrical shapes (for which reason Copernicus said the earth rotates "because" it is spherical, and Gilbert, following the lead, is said to have gone so far as to prove experimentally, at least to his own satisfaction, that a carefully mounted magnetized sphere keeps up a constant rotation). Thus too, Sigmund Freud in *Moses and Monotheism,* after surveying the overwhelmingly unfavorable evidence standing against the central thesis in his book, would say in effect, "But one must not be misled by the evidence."

5.

While developing the position that themata have as legitimate and necessary a place in the pursuit and understanding of science as have observational experience and logical construction, I should make clear that we need not decide now also on the *source* of themata. Our first aim is simply to see their role in science, and to describe some of them, as a folklorist might when he catalogues the traditions and practices of a people. It is not necessary to go further and to make an association of themata with any of the following conceptions: Platonic, Keplerian or Jungian archetypes or images; myths (in the nonderogatory sense, so rarely used in the English language); synthetic a priori knowledge; intuitive apprehension or Galileo's "reason"; a realistic or absolutistic or, for that matter, any other philosophy of science. To show whether any such associations do or do not exist is a task for another time.

I also do not want to imply that the occurrence of themata is characteristic only of science in the last centuries. On the contrary, we see the thematic component at work from the very beginning, in the sources of cosmogonic ideas later found in Hesiod's *Theogony* and in *Genesis*. Indeed, nowhere can one see the persistence of great questions and the obstinacy of certain pre-selected patterns for defining and solving problems better than in cosmologic speculations. The ancient Milesian cosmologic assumptions presented a three-step scheme: At the beginning, in F. M. Cornford's words, there was

> a primal Unity, a state of indistinction or fusion in which factors that will later become distinct are merged together. (2) Out of this Unity there emerge, by separation, parts of opposite things . . . . This sep-

> arating out finally leads to the disposition of the great elemental masses constituting the world-order, and the formation of the heavenly bodies. (3) The Opposites interact or reunite, in meteoric phenomena, or in the production of individual living things . . . .[17]

Now the significant thing to notice is that when we move these conceptions from the animistic to the physical level, this formula of cosmogony recurs point for point, in our day, in the evolutionist camps of modern cosmology. That recent theory of the way the world started proposes a progression of the universe from a mixture of radiation and neutrons at time $t = 0$; through the subsequent stages of differentiation by expansion and neutron decay; and finally to the building up of heavier elements by thermonuclear fusion processes, preparing the ground for the later formation of molecules. And even the ancient main *opposition* to the evolutionary cosmology itself, namely, the tradition of Parmenides, has its equivalent today in the "steady-state" theory of cosmology.

So the questions persist (e.g., concerning the possibility of some "fundamental stuff," of evolution, of structure, of spatial and temporal infinities). And the choices between alternative problem solutions also persist. These thematic continuities indicate the obverse side of the iconoclastic role of science; for science, since its dawn, has also had its more general themata-creating and themata-using function. J. Clark Maxwell expressed this well a century ago in an address on the subject of molecular mechanics:

> The mind of man has perplexed itself with many hard questions. Is space infinite, and in what sense? Is the material world infinite in extent, and are all places

[17] *Principium Sapientiae* (London: Cambridge University Press, 1952), Ch. XI.

> within that extent equally full of matter? Do atoms exist, or is matter infinitely divisible?
>
> The discussion of questions of this kind has been going on ever since man began to reason, and to each of us, as soon as we obtain the use of our faculties, the same old questions arise as fresh as ever. They form as essential a part of science of the nineteenth century of our era, as of that of the fifth century before it.[18]

We may add that thematic questions do not get solved and disposed of. Nineteenth-century atomism triumphs over the ether vortices of Kelvin—but then field theories rise which deal with matter particles again as singularities, now in a twentieth-century-type continuum. The modern version of the cosmological theory based on the thema of a life cycle (Beginning, Evolution, and End) may seem to triumph on experimental grounds over the rival theory based on a thema of Continuous Existence, and throw it out the window—but we can be sure that this thema will come in again through the back door. For contrary to the physical theories in which they find embodiment in x-y terms, themata are not proved or disproved. Rather, they rise and fall and rise again with the tides of contemporaneous usefulness or intellectual fashion. And occasionally a great theme disappears from view, or a new theme develops and struggles to establish itself—at least for a time.

Maxwell's is an unusual concession, but it is not difficult to understand why scientists speak only rarely in such terms. One must not loose sight of the obvious fact that science itself has grown strong because its practitioners have seen how to project their discourse into the x-y

[18] Quoted in C. C. Gillispie, *The Edge of Objectivity* (Princeton, New Jersey: Princeton University Press, 1960), p. 477.

plane. This is the plane of public science,[19] of fairly clear conscious formulations. Here a measure of public agreement is in principle easy to obtain, so that scientists can fruitfully co-operate or disagree with one another, can build on the work of their predecessors, and can teach more or less unambiguously the current content and problems of the field. All fields which claim or pretend to be scientific try similarly to project their concepts, statements, and problems into the x-y plane, to emphasize the phenomenic and analytic-heuristic aspects.

But it is clear that while there can be automatic factories run by means of programmed computers and the feedback from sensing elements, there can be no automatic laboratory. The essence of the automaton is its success in the x-y plane at the expense of the z-direction; (hence automata do not make qualitatively new findings). And the essence of the genial contributor to science is often exactly the opposite—sensitivity in the z-direction even at the expense of success in the x-y plane. For while the x-dimension is never absent even in the most exact of the sciences as pursued by actual persons, it is a direction in which most of us must move *without* explicit or conscious formulation and without training; it is the direction in which the subject matter and the media for communication are entirely different from those invented specifically for discussion of matters in the x-y plane with which the scientist after long training can feel at home.

Therefore it is difficult to find people who are bilingual in this sense. I am not surprised that for most contemporary

[19] For the distinction between public and private science, see G. Holton, "On the Duality and Growth of Science," *American Scientist,* **41** (1953), pp. 89–99.

scientists any discussion which tries to move self-consciously away from the x-y plane is out of bounds. However, it is significant that even in our time the men of genius—such as Einstein, Bohr, Pauli, Born, Schrödinger, Heisenberg—have felt it to be necessary and important to try just that. For the others, for the major body of scientists, the plane of discourse has been progressively tilted or projected from xyz space into the x-y plane. (Perhaps prompted by this example, the same thing is happening more and more in other fields of scholarship.) The themata actually used in science are now largely left implicit rather than explicit. But they are no less important. To understand fully the role a hypothesis or a law has in the development of science we need to see it also as an exemplification of persistent motifs, for example the thema of "constancy" or of "conservation"; of quantification; of atomistic discreteness; of inherently probabilistic behavior; or—to return to our example from Newton—of the interpenetration of the worlds of theology and of physics. Indeed, in this way we can make a useful differentiation that to my knowledge has not been noted before, namely that Newton's *public, experimental,* and *mathematical* philosophy is science carried on in the x-y plane, whereas Newton's more covert and more general *natural* philosophy is science in the x-y-z proposition space.[20]

[20] As Newton writes in the General Scholium, "hypotheses, whether metaphysical or physical, whether of occult qualities or mechanical, have no place in *experimental* philosophy." But in the previous paragraph, at the end of a long passage on the properties of the Deity and His evidences through observable nature, Newton writes: "And thus much concerning God; to discourse of whom from the appearances of things, does certainly belong to *Natural* Philosophy." (Emphasis added.)

6.

I have spoken mostly of the physical sciences. I might, with equal or greater advantage, have dealt with the newer sciences, which do not have a highly developed corpus either of phenomena or of logical calculi and rational structures. In those cases, the z-elements are not only still relatively more prominent but also are discussed with much greater freedom—possibly because at its early stage a field of study still bears the overwhelming imprint of one or a few men of genius. It is they who, I believe, are particularly "themata-prone," and who have the necessary courage (or folly?) to make decisions on thematic grounds.

This was certainly the case in early mechanics and chemistry, and again with relativity and the new quantum mechanics. I suspect that an analogous situation has held in early modern psychology and sociology. Moreover, in those fields, as in the natural sciences during a stage of transformation, the significance and impact of themata is indicated by the fact that they force upon people notions that are usually regarded as paradoxical, ridiculous, or outrageous. I am thinking here of the "absurdities" of Copernicus' moving earth, Bruno's infinite worlds, Galileo's inertial motion of bodies on a horizontal plane, Newton's gravitational action without a palpable medium of communication, Darwin's descent of man from lower creatures, Einstein's twin paradox and maximum speed for signals, Freud's conception of sexuality of children, or Heisenberg's indeterminacy conception. The wide interest and intensity of such debates, among both scientists and enraged or intrigued laymen, is an indication of the strength with which themata—and frequently conflicting ones—are always active in our consciousness.

And the thematic component is most obvious when a science *is* young, and therefore has not yet elaborated the complex hierarchical structure of hypotheses which Braithwaite has pointed out to be the mark of an advanced science. As a result, the chain leading from observational "facts" to the most general hypotheses—those with a large thematic component—is not long, as in, say, modern physics or chemistry, but is fairly short. A physical scientist is used to having his most general and most thematic hypotheses safely out of sight, behind the clouds of a majestic Olympus; and so he is apt to smile when he sees that the altar of other gods stands on such short legs. When for example a chemist interprets a half-dozen clicks on a Geiger counter as the existence of a new chemical element at the end of the Periodic Table, he implicitly (and, if challenged, explicitly) runs up on a ladder of hierarchically connected hypotheses, each of which has *some* demonstrable phenomenic and heuristic-analytic component, until at the top he comes up to the general thematic hypotheses—which he is, by agreement of this fraternity, exempt from going into—namely, the thematic hypotheses of atomicity, of constancy, of the transformability of qualities, of the ordering role of integers. In constrast, the early psychoanalysts, for example, tried to go by a relatively untortuous route from the detail of observed behavior to the generality of powerful principles. Freud himself once warned of the "bad habit of psychoanalysis . . . to take trivia as evidence when they also admit of another, less deep explanatory scheme."[21]

[21] Sigmund Freud, "Ein religiöses Erlebnis," *Imago* XIV (1928), 9.

I do not, of course, say this to condemn a science, but on the contrary to point out a difference between it and the physical sciences which I hope, may help to explain the attitude of "hard" scientists to fields outside their own (or even of psychologists of one school to those of another). At the same time, it may help to elucidate why disciplines such as psychology (and certainly history) are so constructed that they are wrong to imitate the habit in the modern physical sciences to depress or project the discussion forcibly to the x-y plane. When the thematic component is as strong and as explicitly needed as it is in these fields, the criteria of verification should be able to remain explicitly in three-dimensional proposition space. To cite an instance, I am by no means impressed with the "Conclusion" at the end of R. G. Collingwood's influential book, *Essay on Philosophical Method:*[22]

> The natural scientist, beginning with the assumption that nature is rational, has not allowed himself to be turned from that assumption by any of the difficulties into which it has led him; and it is because he has regarded that assumption as not only legitimate but obligatory that he has won the respect of the whole world. If the scientist is obliged to assume that nature is rational, and that any failure to make sense of it is a failure to understand it, the corresponding assumption is obligatory for the historian, and this not least when he is the historian of thought.

This is a statement of the most dangerous kind, not because it is so easy to show it is wrong, but because it is so difficult to show this.

[22] (Oxford: Clarendon Press, 1933), pp. 225–26.

7.

Much could, and should, be said about other problems in the thematic analysis of science, such as the mechanisms by which themata change; or the way in which the choice of a thematic hypothesis governs what we are to look for in the x-y plane and what we do with the findings; or the remarkably small number of different themata that, over time, seem to have played the important roles in the development of science; or the fact, implied in the examples given, that most and perhaps all of these themata are not restricted merely to uses in scientific context, but seem to come from the less specialized ground of our general imaginative capacity.

But in closing I might best simply restate how these conceptions can help us to a view that goes beyond the usual antithetical juxtaposition between science and the humanities. For the much lamented separation between science and the other components of our culture depends on the oversimplification that science is done only in the contingent plane, whereas scholarly or artistic work involves basic decisions of a different kind, with predominantly esthetic, qualitative, mythic elements. In my view this dichotomy is much attenuated, if not eliminated, if we see that in science, too, the contingent plane is not enough, and never has been.

It is surely unnecessary to warn that despite the appearance and reappearance of the same thematic elements in science and outside, we shall not make the mistake of thinking that science and non-science are at bottom somehow the same activity. There are differences which we should treasure. As Whitehead once said about the necessity to tolerate, no, to *welcome* national differences: "Men require of their neighbors something sufficiently akin to be understood, something sufficiently *different* to provoke attention, and something great enough to command admiration." It is in the same sense that we should be prepared to understand both the separateness that gives identity to the study of each field, as well as the kinship that does exist between them.

To return, therefore, to Newton's fifth rule of reasoning: he surely must have known that he could not publish it and remain true to his own work and that of most major innovators. As Newton's suppressed rule stands, it ends, you will recall, with the words: "Those things which neither can be demonstrated from the phenomena nor follow from them by an argument of induction, I hold as hypotheses." To be justified in publishing this rule Newton would have had to add something—perhaps this sentence: "And such hypotheses, namely thematic hypotheses, do also have a place in natural philosophy."

Index